Table of contents

Section 1

Find the sum

Complete all the activities (Addition).

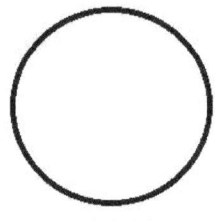

SCORE

1.	600 + 191	2.	981 + 644	3.	700 + 938	4.	207 + 900	5.	372 + 367	6.	959 + 819	7.	489 + 460

8.	186 + 866	9.	856 + 542	10.	214 + 199	11.	463 + 482	12.	776 + 535	13.	816 + 147	14.	497 + 835

15.	885 + 647	16.	436 + 127	17.	168 + 153	18.	407 + 196	19.	340 + 494	20.	360 + 596	21.	630 + 372

22.	324 + 791	23.	457 + 437	24.	132 + 851	25.	478 + 851	26.	371 + 831	27.	742 + 138	28.	163 + 218

29.	414 + 546	30.	827 + 652	31.	542 + 371	32.	115 + 724	33.	930 + 932	34.	751 + 437	35.	153 + 989

36.	626 + 462	37.	282 + 616	38.	568 + 191	39.	258 + 984	40.	179 + 479	41.	115 + 645	42.	457 + 135

Name:................................. Date:.................................

Find the sum

Complete all the activities (Addition).

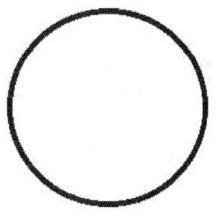

SCORE

1. 600 + 191 791	2. 981 + 644 1 625	3. 700 + 938 1 638	4. 207 + 900 1 107	5. 372 + 367 739	6. 959 + 819 1 778	7. 489 + 460 949
8. 186 + 866 1 052	9. 856 + 542 1 398	10. 214 + 199 413	11. 463 + 482 945	12. 776 + 535 1 311	13. 816 + 147 963	14. 497 + 835 1 332
15. 885 + 647 1 532	16. 436 + 127 563	17. 168 + 153 321	18. 407 + 196 603	19. 340 + 494 834	20. 360 + 596 956	21. 630 + 372 1 002
22. 324 + 791 1 115	23. 457 + 437 894	24. 132 + 851 983	25. 478 + 851 1 329	26. 371 + 831 1 202	27. 742 + 138 880	28. 163 + 218 381
29. 414 + 546 960	30. 827 + 652 1 479	31. 542 + 371 913	32. 115 + 724 839	33. 930 + 932 1 862	34. 751 + 437 1 188	35. 153 + 989 1 142
36. 626 + 462 1 088	37. 282 + 616 898	38. 568 + 191 759	39. 258 + 984 1 242	40. 179 + 479 658	41. 115 + 645 760	42. 457 + 135 592

Name:............................   Date:............................

Find the sum

Complete all the activities (Addition).

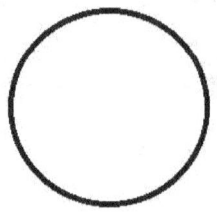

SCORE

1. 665
   + 773
   ----

2. 891
   + 843
   ----

3. 907
   + 819
   ----

4. 941
   + 865
   ----

5. 321
   + 960
   ----

6. 691
   + 548
   ----

7. 326
   + 581
   ----

8. 759
   + 249
   ----

9. 688
   + 255
   ----

10. 883
    + 833
    ----

11. 650
    + 553
    ----

12. 600
    + 808
    ----

13. 582
    + 922
    ----

14. 561
    + 276
    ----

15. 758
    + 990
    ----

16. 530
    + 944
    ----

17. 318
    + 525
    ----

18. 157
    + 965
    ----

19. 114
    + 283
    ----

20. 195
    + 534
    ----

21. 209
    + 742
    ----

22. 408
    + 375
    ----

23. 732
    + 821
    ----

24. 185
    + 781
    ----

25. 695
    + 797
    ----

26. 254
    + 819
    ----

27. 310
    + 383
    ----

28. 816
    + 212
    ----

29. 272
    + 183
    ----

30. 362
    + 984
    ----

31. 357
    + 765
    ----

32. 207
    + 330
    ----

33. 530
    + 746
    ----

34. 330
    + 800
    ----

35. 524
    + 794
    ----

36. 235
    + 505
    ----

37. 159
    + 449
    ----

38. 723
    + 907
    ----

39. 245
    + 326
    ----

40. 176
    + 760
    ----

41. 838
    + 779
    ----

42. 580
    + 960

Name:.................................   Date:................................

Find the sum

Complete all the activities (Addition).

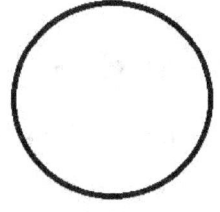

SCORE

1.	665	2.	891	3.	907	4.	941	5.	321	6.	691	7.	326
	+ 773		+ 843		+ 819		+ 865		+ 960		+ 548		+ 581
	1 438		1 734		1 726		1 806		1 281		1 239		907

8.	759	9.	688	10.	883	11.	650	12.	600	13.	582	14.	561
	+ 249		+ 255		+ 833		+ 553		+ 808		+ 922		+ 276
	1 008		943		1 716		1 203		1 408		1 504		837

15.	758	16.	530	17.	318	18.	157	19.	114	20.	195	21.	209
	+ 990		+ 944		+ 525		+ 965		+ 283		+ 534		+ 742
	1 748		1 474		843		1 122		397		729		951

22.	408	23.	732	24.	185	25.	695	26.	254	27.	310	28.	816
	+ 375		+ 821		+ 781		+ 797		+ 819		+ 383		+ 212
	783		1 553		966		1 492		1 073		693		1 028

29.	272	30.	362	31.	357	32.	207	33.	530	34.	330	35.	524
	+ 183		+ 984		+ 765		+ 330		+ 746		+ 800		+ 794
	455		1 346		1 122		537		1 276		1 130		1 318

36.	235	37.	159	38.	723	39.	245	40.	176	41.	838	42.	580
	+ 505		+ 449		+ 907		+ 326		+ 760		+ 779		+ 960
	740		608		1 630		571		936		1 617		1 540

Name:................................   Date:................................

Find the sum

Complete all the activities (Addition).

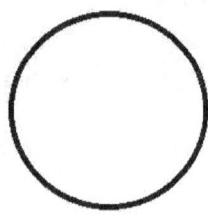

SCORE

1. 835 + 710	2. 734 + 934	3. 708 + 944	4. 792 + 930	5. 920 + 211	6. 276 + 443	7. 635 + 178
8. 844 + 985	9. 590 + 384	10. 357 + 900	11. 662 + 748	12. 760 + 859	13. 659 + 909	14. 966 + 946
15. 107 + 601	16. 246 + 577	17. 285 + 246	18. 494 + 819	19. 112 + 396	20. 192 + 579	21. 102 + 304
22. 584 + 948	23. 273 + 380	24. 372 + 167	25. 481 + 190	26. 269 + 554	27. 522 + 679	28. 398 + 187
29. 949 + 874	30. 375 + 798	31. 977 + 846	32. 923 + 419	33. 964 + 698	34. 258 + 489	35. 683 + 125
36. 972 + 295	37. 292 + 777	38. 898 + 458	39. 264 + 811	40. 687 + 379	41. 637 + 900	42. 565 + 916

Name:................................ Date:................................

Find the sum

Complete all the activities (Addition).

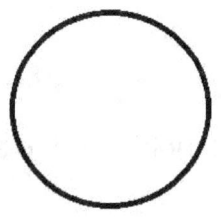

SCORE

1.	835 + 710 1 545	2.	734 + 934 1 668	3.	708 + 944 1 652	4.	792 + 930 1 722	5.	920 + 211 1 131	6.	276 + 443 719	7.	635 + 178 813
8.	844 + 985 1 829	9.	590 + 384 974	10.	357 + 900 1 257	11.	662 + 748 1 410	12.	760 + 859 1 619	13.	659 + 909 1 568	14.	966 + 946 1 912
15.	107 + 601 708	16.	246 + 577 823	17.	285 + 246 531	18.	494 + 819 1 313	19.	112 + 396 508	20.	192 + 579 771	21.	102 + 304 406
22.	584 + 948 1 532	23.	273 + 380 653	24.	372 + 167 539	25.	481 + 190 671	26.	269 + 554 823	27.	522 + 679 1 201	28.	398 + 187 585
29.	949 + 874 1 823	30.	375 + 798 1 173	31.	977 + 846 1 823	32.	923 + 419 1 342	33.	964 + 698 1 662	34.	258 + 489 747	35.	683 + 125 808
36.	972 + 295 1 267	37.	292 + 777 1 069	38.	898 + 458 1 356	39.	264 + 811 1 075	40.	687 + 379 1 066	41.	637 + 900 1 537	42.	565 + 916 1 481

Name:................................ Date:................................

Find the sum

Complete all the activities (Addition).

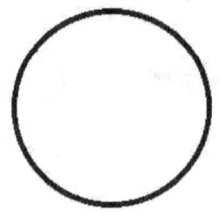

1. 432 + 691	2. 155 + 605	3. 847 + 406	4. 868 + 238	5. 121 + 393	6. 601 + 255	7. 379 + 266
8. 983 + 574	9. 222 + 882	10. 482 + 456	11. 217 + 974	12. 362 + 744	13. 779 + 939	14. 923 + 238
15. 287 + 995	16. 261 + 177	17. 381 + 108	18. 986 + 373	19. 643 + 652	20. 641 + 377	21. 256 + 835
22. 274 + 898	23. 277 + 743	24. 179 + 997	25. 827 + 213	26. 257 + 754	27. 517 + 209	28. 799 + 457
29. 429 + 410	30. 825 + 907	31. 482 + 730	32. 950 + 125	33. 232 + 523	34. 873 + 372	35. 279 + 710
36. 929 + 568	37. 341 + 205	38. 300 + 381	39. 800 + 463	40. 567 + 136	41. 860 + 925	42. 410 + 174

Find the sum

Name:.............................   Date:...............................

Complete all the activities (Addition).

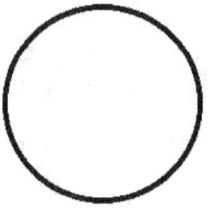

SCORE

1. 432 + 691 1 123	2. 155 + 605 760	3. 847 + 406 1 253	4. 868 + 238 1 106	5. 121 + 393 514	6. 601 + 255 856	7. 379 + 266 645

8. 983 + 574 1 557	9. 222 + 882 1 104	10. 482 + 456 938	11. 217 + 974 1 191	12. 362 + 744 1 106	13. 779 + 939 1 718	14. 923 + 238 1 161

15. 287 + 995 1 282	16. 261 + 177 438	17. 381 + 108 489	18. 986 + 373 1 359	19. 643 + 652 1 295	20. 641 + 377 1 018	21. 256 + 835 1 091

22. 274 + 898 1 172	23. 277 + 743 1 020	24. 179 + 997 1 176	25. 827 + 213 1 040	26. 257 + 754 1 011	27. 517 + 209 726	28. 799 + 457 1 256

29. 429 + 410 839	30. 825 + 907 1 732	31. 482 + 730 1 212	32. 950 + 125 1 075	33. 232 + 523 755	34. 873 + 372 1 245	35. 279 + 710 989

36. 929 + 568 1 497	37. 341 + 205 546	38. 300 + 381 681	39. 800 + 463 1 263	40. 567 + 136 703	41. 860 + 925 1 785	42. 410 + 174 584

Name:................................ Date:............................

Find the sum

Complete all the activities (Addition).

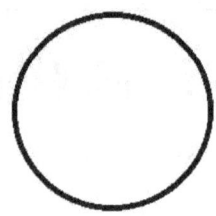

SCORE

1. 837 + 507	2. 659 + 272	3. 308 + 628	4. 943 + 333	5. 826 + 734	6. 894 + 524	7. 884 + 248
8. 766 + 752	9. 869 + 806	10. 818 + 218	11. 283 + 688	12. 513 + 412	13. 907 + 518	14. 656 + 966
15. 843 + 789	16. 146 + 557	17. 420 + 876	18. 287 + 724	19. 864 + 701	20. 164 + 649	21. 766 + 697
22. 741 + 606	23. 310 + 775	24. 496 + 652	25. 695 + 814	26. 266 + 154	27. 207 + 292	28. 890 + 249
29. 993 + 346	30. 364 + 789	31. 971 + 976	32. 489 + 138	33. 687 + 437	34. 204 + 108	35. 960 + 143
36. 888 + 268	37. 742 + 269	38. 451 + 585	39. 946 + 906	40. 202 + 232	41. 122 + 104	42. 370 + 129

Find the sum

Complete all the activities (Addition).

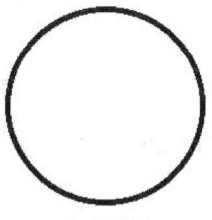

SCORE

1.	837 + 507 1 344	2.	659 + 272 931	3.	308 + 628 936	4.	943 + 333 1 276	5.	826 + 734 1 560	6.	894 + 524 1 418	7.	884 + 248 1 132

| 8. | 766
+ 752
1 518 | 9. | 869
+ 806
1 675 | 10. | 818
+ 218
1 036 | 11. | 283
+ 688
971 | 12. | 513
+ 412
925 | 13. | 907
+ 518
1 425 | 14. | 656
+ 966
1 622 |

| 15. | 843
+ 789
1 632 | 16. | 146
+ 557
703 | 17. | 420
+ 876
1 296 | 18. | 287
+ 724
1 011 | 19. | 864
+ 701
1 565 | 20. | 164
+ 649
813 | 21. | 766
+ 697
1 463 |

| 22. | 741
+ 606
1 347 | 23. | 310
+ 775
1 085 | 24. | 496
+ 652
1 148 | 25. | 695
+ 814
1 509 | 26. | 266
+ 154
420 | 27. | 207
+ 292
499 | 28. | 890
+ 249
1 139 |

| 29. | 993
+ 346
1 339 | 30. | 364
+ 789
1 153 | 31. | 971
+ 976
1 947 | 32. | 489
+ 138
627 | 33. | 687
+ 437
1 124 | 34. | 204
+ 108
312 | 35. | 960
+ 143
1 103 |

| 36. | 888
+ 268
1 156 | 37. | 742
+ 269
1 011 | 38. | 451
+ 585
1 036 | 39. | 946
+ 906
1 852 | 40. | 202
+ 232
434 | 41. | 122
+ 104
226 | 42. | 370
+ 129
499 |

Find the sum

Complete all the activities (Addition).

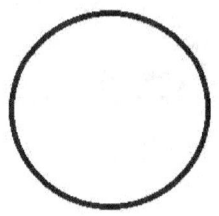

SCORE

1.   622
    + 991
	2. 450
	    + 228
		3. 985
		    + 138
			4. 730
			    + 617
				5. 437
				    + 374
					6. 677
					    + 789
						7. 307
						    + 703

8.   918
    + 872
	9. 671
	    + 487
		10. 691
		    + 653
			11. 925
			    + 997
				12. 412
				    + 937
					13. 840
					    + 405
						14. 302
						    + 187

15.   244
    + 232
	16. 767
	    + 967
		17. 706
		    + 460
			18. 515
			    + 696
				19. 986
				    + 975
					20. 122
					    + 922
						21. 412
						    + 370

22.   831
    + 467
	23. 591
	    + 661
		24. 378
		    + 343
			25. 987
			    + 329
				26. 561
				    + 634
					27. 413
					    + 717
						28. 203
						    + 624

29.   679
    + 512
	30. 387
	    + 680
		31. 834
		    + 642
			32. 473
			    + 305
				33. 607
				    + 709
					34. 183
					    + 846
						35. 289
						    + 322

36.   885
    + 694
	37. 465
	    + 767
		38. 341
		    + 655
			39. 266
			    + 943
				40. 447
				    + 279
					41. 127
					    + 959
						42. 199
						    + 914

Name:................................... Date:...................................

Find the sum
Complete all the activities (Addition).

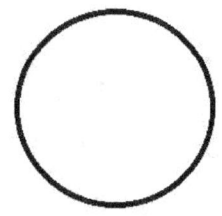

1. 622 + 991 1 613	2. 450 + 228 678	3. 985 + 138 1 123	4. 730 + 617 1 347	5. 437 + 374 811	6. 677 + 789 1 466	7. 307 + 703 1 010
8. 918 + 872 1 790	9. 671 + 487 1 158	10. 691 + 653 1 344	11. 925 + 997 1 922	12. 412 + 937 1 349	13. 840 + 405 1 245	14. 302 + 187 489
15. 244 + 232 476	16. 767 + 967 1 734	17. 706 + 460 1 166	18. 515 + 696 1 211	19. 986 + 975 1 961	20. 122 + 922 1 044	21. 412 + 370 782
22. 831 + 467 1 298	23. 591 + 661 1 252	24. 378 + 343 721	25. 987 + 329 1 316	26. 561 + 634 1 195	27. 413 + 717 1 130	28. 203 + 624 827
29. 679 + 512 1 191	30. 387 + 680 1 067	31. 834 + 642 1 476	32. 473 + 305 778	33. 607 + 709 1 316	34. 183 + 846 1 029	35. 289 + 322 611
36. 885 + 694 1 579	37. 465 + 767 1 232	38. 341 + 655 996	39. 266 + 943 1 209	40. 447 + 279 726	41. 127 + 959 1 086	42. 199 + 914 1 113

Name:................................ Date:................................

Find the sum

Complete all the activities (Addition).

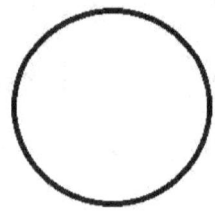

1. 670 + 233	2. 654 + 988	3. 357 + 889	4. 116 + 574	5. 689 + 441	6. 188 + 769	7. 896 + 626
8. 860 + 881	9. 578 + 531	10. 154 + 765	11. 443 + 865	12. 269 + 645	13. 476 + 547	14. 458 + 735
15. 165 + 208	16. 653 + 875	17. 543 + 558	18. 633 + 474	19. 155 + 193	20. 589 + 398	21. 306 + 411
22. 106 + 124	23. 679 + 571	24. 757 + 639	25. 183 + 565	26. 700 + 595	27. 905 + 433	28. 570 + 804
29. 963 + 110	30. 613 + 809	31. 532 + 380	32. 509 + 581	33. 787 + 312	34. 325 + 438	35. 992 + 486
36. 826 + 485	37. 941 + 754	38. 559 + 133	39. 860 + 245	40. 965 + 904	41. 789 + 817	42. 652 + 599

Find the sum

Complete all the activities (Addition).

Name:.............................. Date:..............................

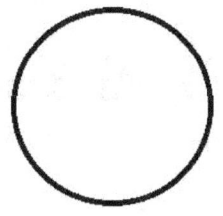

SCORE

| 1. | 670
+ 233
903 | 2. | 654
+ 988
1 642 | 3. | 357
+ 889
1 246 | 4. | 116
+ 574
690 | 5. | 689
+ 441
1 130 | 6. | 188
+ 769
957 | 7. | 896
+ 626
1 522 |

| 8. | 860
+ 881
1 741 | 9. | 578
+ 531
1 109 | 10. | 154
+ 765
919 | 11. | 443
+ 865
1 308 | 12. | 269
+ 645
914 | 13. | 476
+ 547
1 023 | 14. | 458
+ 735
1 193 |

| 15. | 165
+ 208
373 | 16. | 653
+ 875
1 528 | 17. | 543
+ 558
1 101 | 18. | 633
+ 474
1 107 | 19. | 155
+ 193
348 | 20. | 589
+ 398
987 | 21. | 306
+ 411
717 |

| 22. | 106
+ 124
230 | 23. | 679
+ 571
1 250 | 24. | 757
+ 639
1 396 | 25. | 183
+ 565
748 | 26. | 700
+ 595
1 295 | 27. | 905
+ 433
1 338 | 28. | 570
+ 804
1 374 |

| 29. | 963
+ 110
1 073 | 30. | 613
+ 809
1 422 | 31. | 532
+ 380
912 | 32. | 509
+ 581
1 090 | 33. | 787
+ 312
1 099 | 34. | 325
+ 438
763 | 35. | 992
+ 486
1 478 |

| 36. | 826
+ 485
1 311 | 37. | 941
+ 754
1 695 | 38. | 559
+ 133
692 | 39. | 860
+ 245
1 105 | 40. | 965
+ 904
1 869 | 41. | 789
+ 817
1 606 | 42. | 652
+ 599
1 251 |

Name:......................................   Date:.............................

Find the sum

Complete all the activities (Addition).

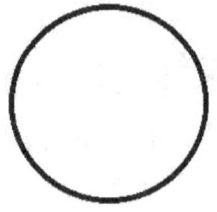

SCORE

1. 507 + 158	2. 969 + 210	3. 987 + 828	4. 510 + 145	5. 580 + 779	6. 866 + 925	7. 901 + 524
8. 335 + 467	9. 346 + 419	10. 647 + 696	11. 738 + 315	12. 653 + 251	13. 873 + 538	14. 943 + 818
15. 144 + 708	16. 120 + 713	17. 727 + 908	18. 174 + 729	19. 343 + 215	20. 450 + 264	21. 975 + 781
22. 904 + 262	23. 888 + 634	24. 231 + 436	25. 909 + 598	26. 658 + 470	27. 271 + 954	28. 860 + 292
29. 504 + 818	30. 336 + 549	31. 884 + 691	32. 101 + 871	33. 759 + 252	34. 297 + 991	35. 449 + 605
36. 224 + 230	37. 968 + 483	38. 720 + 956	39. 410 + 490	40. 362 + 339	41. 407 + 607	42. 908 + 545

Find the sum

Name:.................................   Date:.................................

Complete all the activities (Addition).

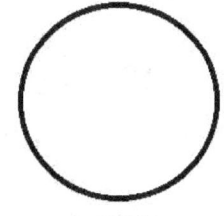

SCORE

1. 507 + 158 665	2. 969 + 210 1 179	3. 987 + 828 1 815	4. 510 + 145 655	5. 580 + 779 1 359	6. 866 + 925 1 791	7. 901 + 524 1 425

8. 335 + 467 802	9. 346 + 419 765	10. 647 + 696 1 343	11. 738 + 315 1 053	12. 653 + 251 904	13. 873 + 538 1 411	14. 943 + 818 1 761

15. 144 + 708 852	16. 120 + 713 833	17. 727 + 908 1 635	18. 174 + 729 903	19. 343 + 215 558	20. 450 + 264 714	21. 975 + 781 1 756

22. 904 + 262 1 166	23. 888 + 634 1 522	24. 231 + 436 667	25. 909 + 598 1 507	26. 658 + 470 1 128	27. 271 + 954 1 225	28. 860 + 292 1 152

29. 504 + 818 1 322	30. 336 + 549 885	31. 884 + 691 1 575	32. 101 + 871 972	33. 759 + 252 1 011	34. 297 + 991 1 288	35. 449 + 605 1 054

36. 224 + 230 454	37. 968 + 483 1 451	38. 720 + 956 1 676	39. 410 + 490 900	40. 362 + 339 701	41. 407 + 607 1 014	42. 908 + 545 1 453

Name:..................................   Date:..............................

Find the sum

Complete all the activities (Addition).

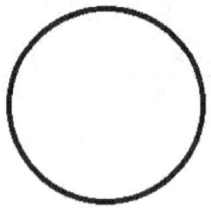

SCORE

1. 150 + 649	2. 543 + 772	3. 148 + 881	4. 325 + 523	5. 878 + 900	6. 379 + 384	7. 781 + 133
8. 503 + 736	9. 436 + 239	10. 986 + 404	11. 474 + 527	12. 501 + 787	13. 334 + 903	14. 881 + 114
15. 556 + 996	16. 243 + 377	17. 782 + 554	18. 821 + 210	19. 708 + 525	20. 943 + 567	21. 380 + 700
22. 351 + 167	23. 204 + 436	24. 275 + 377	25. 987 + 625	26. 902 + 268	27. 620 + 604	28. 172 + 622
29. 247 + 755	30. 538 + 244	31. 999 + 311	32. 966 + 881	33. 758 + 101	34. 601 + 326	35. 311 + 393
36. 127 + 910	37. 294 + 519	38. 124 + 773	39. 276 + 434	40. 682 + 898	41. 728 + 604	42. 126 + 930

Find the sum

Complete all the activities (Addition).

Name:................................   Date:.................................

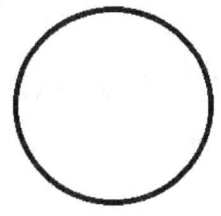

SCORE

1. 150 + 649 799	2. 543 + 772 1 315	3. 148 + 881 1 029	4. 325 + 523 848	5. 878 + 900 1 778	6. 379 + 384 763	7. 781 + 133 914
8. 503 + 736 1 239	9. 436 + 239 675	10. 986 + 404 1 390	11. 474 + 527 1 001	12. 501 + 787 1 288	13. 334 + 903 1 237	14. 881 + 114 995
15. 556 + 996 1 552	16. 243 + 377 620	17. 782 + 554 1 336	18. 821 + 210 1 031	19. 708 + 525 1 233	20. 943 + 567 1 510	21. 380 + 700 1 080
22. 351 + 167 518	23. 204 + 436 640	24. 275 + 377 652	25. 987 + 625 1 612	26. 902 + 268 1 170	27. 620 + 604 1 224	28. 172 + 622 794
29. 247 + 755 1 002	30. 538 + 244 782	31. 999 + 311 1 310	32. 966 + 881 1 847	33. 758 + 101 859	34. 601 + 326 927	35. 311 + 393 704
36. 127 + 910 1 037	37. 294 + 519 813	38. 124 + 773 897	39. 276 + 434 710	40. 682 + 898 1 580	41. 728 + 604 1 332	42. 126 + 930 1 056

Name:.................................... Date:....................................

Find the sum

Complete all the activities (Addition).

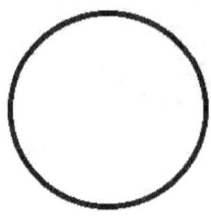

SCORE

1. 194 + 974	2. 756 + 279	3. 483 + 299	4. 962 + 952	5. 986 + 799	6. 567 + 168	7. 733 + 469
8. 408 + 203	9. 834 + 948	10. 165 + 694	11. 598 + 991	12. 972 + 201	13. 579 + 398	14. 446 + 812
15. 173 + 639	16. 515 + 494	17. 286 + 136	18. 787 + 783	19. 344 + 423	20. 843 + 113	21. 430 + 174
22. 923 + 494	23. 951 + 955	24. 140 + 806	25. 126 + 579	26. 164 + 779	27. 539 + 431	28. 432 + 202
29. 101 + 655	30. 698 + 185	31. 852 + 249	32. 880 + 447	33. 384 + 173	34. 882 + 697	35. 681 + 624
36. 858 + 511	37. 205 + 153	38. 364 + 525	39. 167 + 579	40. 250 + 623	41. 357 + 639	42. 787 + 592

Find the sum

Name:.............................   Date:..............................

Complete all the activities (Addition).

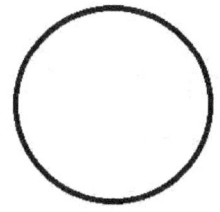

SCORE

1. 194 + 974 1 168	2. 756 + 279 1 035	3. 483 + 299 782	4. 962 + 952 1 914	5. 986 + 799 1 785	6. 567 + 168 735	7. 733 + 469 1 202
8. 408 + 203 611	9. 834 + 948 1 782	10. 165 + 694 859	11. 598 + 991 1 589	12. 972 + 201 1 173	13. 579 + 398 977	14. 446 + 812 1 258
15. 173 + 639 812	16. 515 + 494 1 009	17. 286 + 136 422	18. 787 + 783 1 570	19. 344 + 423 767	20. 843 + 113 956	21. 430 + 174 604
22. 923 + 494 1 417	23. 951 + 955 1 906	24. 140 + 806 946	25. 126 + 579 705	26. 164 + 779 943	27. 539 + 431 970	28. 432 + 202 634
29. 101 + 655 756	30. 698 + 185 883	31. 852 + 249 1 101	32. 880 + 447 1 327	33. 384 + 173 557	34. 882 + 697 1 579	35. 681 + 624 1 305
36. 858 + 511 1 369	37. 205 + 153 358	38. 364 + 525 889	39. 167 + 579 746	40. 250 + 623 873	41. 357 + 639 996	42. 787 + 592 1 379

Name:................................ Date:................................

Find the sum

Complete all the activities (Addition).

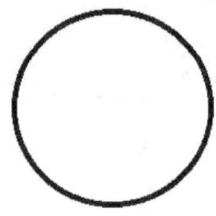

SCORE

1. 139 + 518	2. 458 + 550	3. 190 + 349	4. 476 + 327	5. 103 + 876	6. 989 + 320	7. 393 + 933
8. 498 + 816	9. 621 + 251	10. 264 + 831	11. 980 + 704	12. 587 + 308	13. 553 + 174	14. 971 + 278
15. 547 + 207	16. 222 + 508	17. 440 + 231	18. 467 + 301	19. 804 + 440	20. 647 + 266	21. 202 + 618
22. 529 + 923	23. 401 + 933	24. 952 + 253	25. 278 + 624	26. 675 + 894	27. 179 + 914	28. 925 + 434
29. 930 + 250	30. 996 + 116	31. 377 + 298	32. 779 + 895	33. 754 + 580	34. 879 + 262	35. 565 + 297
36. 112 + 281	37. 603 + 997	38. 728 + 462	39. 275 + 664	40. 906 + 619	41. 221 + 818	42. 100 + 922

Name:..................................... Date:.....................................

Find the sum

Complete all the activities (Addition).

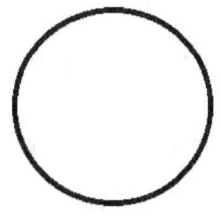

SCORE

| 1. | 139
+ 518
657 | 2. | 458
+ 550
1 008 | 3. | 190
+ 349
539 | 4. | 476
+ 327
803 | 5. | 103
+ 876
979 | 6. | 989
+ 320
1 309 | 7. | 393
+ 933
1 326 |

| 8. | 498
+ 816
1 314 | 9. | 621
+ 251
872 | 10. | 264
+ 831
1 095 | 11. | 980
+ 704
1 684 | 12. | 587
+ 308
895 | 13. | 553
+ 174
727 | 14. | 971
+ 278
1 249 |

| 15. | 547
+ 207
754 | 16. | 222
+ 508
730 | 17. | 440
+ 231
671 | 18. | 467
+ 301
768 | 19. | 804
+ 440
1 244 | 20. | 647
+ 266
913 | 21. | 202
+ 618
820 |

| 22. | 529
+ 923
1 452 | 23. | 401
+ 933
1 334 | 24. | 952
+ 253
1 205 | 25. | 278
+ 624
902 | 26. | 675
+ 894
1 569 | 27. | 179
+ 914
1 093 | 28. | 925
+ 434
1 359 |

| 29. | 930
+ 250
1 180 | 30. | 996
+ 116
1 112 | 31. | 377
+ 298
675 | 32. | 779
+ 895
1 674 | 33. | 754
+ 580
1 334 | 34. | 879
+ 262
1 141 | 35. | 565
+ 297
862 |

| 36. | 112
+ 281
393 | 37. | 603
+ 997
1 600 | 38. | 728
+ 462
1 190 | 39. | 275
+ 664
939 | 40. | 906
+ 619
1 525 | 41. | 221
+ 818
1 039 | 42. | 100
+ 922
1 022 |

Find the sum

Complete all the activities (Addition).

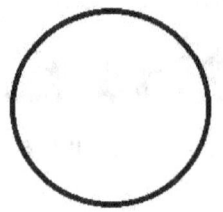

SCORE

1. 531 + 327	2. 429 + 525	3. 306 + 134	4. 370 + 654	5. 443 + 631	6. 889 + 904	7. 838 + 634
8. 268 + 820	9. 155 + 790	10. 687 + 980	11. 612 + 494	12. 391 + 182	13. 286 + 562	14. 259 + 366
15. 343 + 477	16. 894 + 558	17. 297 + 857	18. 464 + 528	19. 251 + 453	20. 955 + 532	21. 926 + 457
22. 837 + 972	23. 208 + 824	24. 624 + 640	25. 495 + 186	26. 564 + 429	27. 231 + 120	28. 409 + 137
29. 934 + 348	30. 339 + 743	31. 874 + 104	32. 411 + 965	33. 735 + 431	34. 706 + 980	35. 269 + 740
36. 669 + 700	37. 835 + 255	38. 153 + 256	39. 656 + 419	40. 972 + 461	41. 938 + 617	42. 870 + 258

Name:.................................   Date:................................

Find the sum

Complete all the activities (Addition).

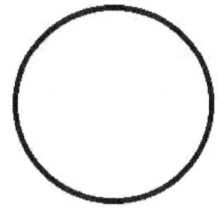

SCORE

1. 531 + 327 858	2. 429 + 525 954	3. 306 + 134 440	4. 370 + 654 1 024	5. 443 + 631 1 074	6. 889 + 904 1 793	7. 838 + 634 1 472
8. 268 + 820 1 088	9. 155 + 790 945	10. 687 + 980 1 667	11. 612 + 494 1 106	12. 391 + 182 573	13. 286 + 562 848	14. 259 + 366 625
15. 343 + 477 820	16. 894 + 558 1 452	17. 297 + 857 1 154	18. 464 + 528 992	19. 251 + 453 704	20. 955 + 532 1 487	21. 926 + 457 1 383
22. 837 + 972 1 809	23. 208 + 824 1 032	24. 624 + 640 1 264	25. 495 + 186 681	26. 564 + 429 993	27. 231 + 120 351	28. 409 + 137 546
29. 934 + 348 1 282	30. 339 + 743 1 082	31. 874 + 104 978	32. 411 + 965 1 376	33. 735 + 431 1 166	34. 706 + 980 1 686	35. 269 + 740 1 009
36. 669 + 700 1 369	37. 835 + 255 1 090	38. 153 + 256 409	39. 656 + 419 1 075	40. 972 + 461 1 433	41. 938 + 617 1 555	42. 870 + 258 1 128

Find the sum

Complete all the activities (Addition).

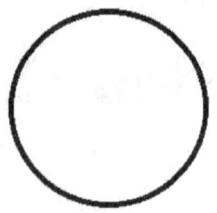

SCORE

| 1. | 912
+ 565 | 2. | 427
+ 278 | 3. | 547
+ 427 | 4. | 592
+ 902 | 5. | 214
+ 312 | 6. | 657
+ 923 | 7. | 146
+ 871 |

| 8. | 336
+ 777 | 9. | 218
+ 643 | 10. | 471
+ 755 | 11. | 516
+ 863 | 12. | 594
+ 975 | 13. | 965
+ 254 | 14. | 480
+ 616 |

| 15. | 374
+ 878 | 16. | 804
+ 357 | 17. | 233
+ 253 | 18. | 153
+ 270 | 19. | 104
+ 529 | 20. | 533
+ 841 | 21. | 248
+ 766 |

| 22. | 587
+ 623 | 23. | 597
+ 229 | 24. | 210
+ 783 | 25. | 182
+ 296 | 26. | 343
+ 395 | 27. | 246
+ 215 | 28. | 741
+ 149 |

| 29. | 970
+ 451 | 30. | 311
+ 394 | 31. | 946
+ 976 | 32. | 628
+ 335 | 33. | 507
+ 400 | 34. | 790
+ 892 | 35. | 818
+ 166 |

| 36. | 773
+ 688 | 37. | 210
+ 629 | 38. | 607
+ 749 | 39. | 376
+ 970 | 40. | 478
+ 836 | 41. | 144
+ 696 | 42. | 533
+ 179 |

Find the sum

Complete all the activities (Addition).

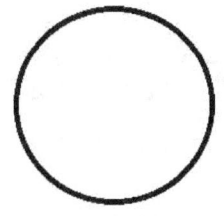

SCORE

1. 912
 + 565
 1 477

2. 427
 + 278
 705

3. 547
 + 427
 974

4. 592
 + 902
 1 494

5. 214
 + 312
 526

6. 657
 + 923
 1 580

7. 146
 + 871
 1 017

8. 336
 + 777
 1 113

9. 218
 + 643
 861

10. 471
 + 755
 1 226

11. 516
 + 863
 1 379

12. 594
 + 975
 1 569

13. 965
 + 254
 1 219

14. 480
 + 616
 1 096

15. 374
 + 878
 1 252

16. 804
 + 357
 1 161

17. 233
 + 253
 486

18. 153
 + 270
 423

19. 104
 + 529
 633

20. 533
 + 841
 1 374

21. 248
 + 766
 1 014

22. 587
 + 623
 1 210

23. 597
 + 229
 826

24. 210
 + 783
 993

25. 182
 + 296
 478

26. 343
 + 395
 738

27. 246
 + 215
 461

28. 741
 + 149
 890

29. 970
 + 451
 1 421

30. 311
 + 394
 705

31. 946
 + 976
 1 922

32. 628
 + 335
 963

33. 507
 + 400
 907

34. 790
 + 892
 1 682

35. 818
 + 166
 984

36. 773
 + 688
 1 461

37. 210
 + 629
 839

38. 607
 + 749
 1 356

39. 376
 + 970
 1 346

40. 478
 + 836
 1 314

41. 144
 + 696
 840

42. 533
 + 179
 712

Name:................................   Date:...........................

Find the sum

Complete all the activities (Addition).

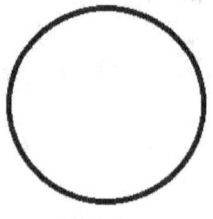

SCORE

1. 634 + 509	2. 210 + 118	3. 511 + 105	4. 299 + 652	5. 317 + 239	6. 894 + 271	7. 238 + 351
8. 215 + 296	9. 438 + 317	10. 965 + 116	11. 522 + 654	12. 285 + 617	13. 607 + 991	14. 772 + 100
15. 972 + 474	16. 350 + 351	17. 542 + 549	18. 700 + 997	19. 290 + 772	20. 480 + 813	21. 867 + 221
22. 471 + 970	23. 734 + 436	24. 612 + 125	25. 855 + 282	26. 498 + 825	27. 730 + 517	28. 656 + 979
29. 323 + 592	30. 509 + 519	31. 332 + 299	32. 680 + 842	33. 888 + 643	34. 728 + 687	35. 650 + 217
36. 106 + 665	37. 836 + 252	38. 248 + 401	39. 283 + 315	40. 882 + 271	41. 479 + 747	42. 244 + 790

Find the sum

Complete all the activities (Addition).

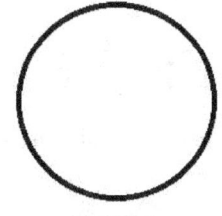

SCORE

1. 634 + 509 1 143	2. 210 + 118 328	3. 511 + 105 616	4. 299 + 652 951	5. 317 + 239 556	6. 894 + 271 1 165	7. 238 + 351 589
8. 215 + 296 511	9. 438 + 317 755	10. 965 + 116 1 081	11. 522 + 654 1 176	12. 285 + 617 902	13. 607 + 991 1 598	14. 772 + 100 872
15. 972 + 474 1 446	16. 350 + 351 701	17. 542 + 549 1 091	18. 700 + 997 1 697	19. 290 + 772 1 062	20. 480 + 813 1 293	21. 867 + 221 1 088
22. 471 + 970 1 441	23. 734 + 436 1 170	24. 612 + 125 737	25. 855 + 282 1 137	26. 498 + 825 1 323	27. 730 + 517 1 247	28. 656 + 979 1 635
29. 323 + 592 915	30. 509 + 519 1 028	31. 332 + 299 631	32. 680 + 842 1 522	33. 888 + 643 1 531	34. 728 + 687 1 415	35. 650 + 217 867
36. 106 + 665 771	37. 836 + 252 1 088	38. 248 + 401 649	39. 283 + 315 598	40. 882 + 271 1 153	41. 479 + 747 1 226	42. 244 + 790 1 034

Name:................................   Date:...............................

Find the sum

Complete all the activities (Addition).

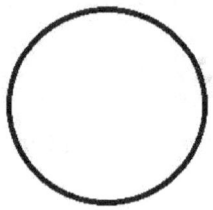

SCORE

1. 976 + 524	2. 620 + 125	3. 892 + 720	4. 774 + 697	5. 363 + 960	6. 581 + 668	7. 540 + 628
8. 182 + 191	9. 707 + 923	10. 473 + 737	11. 276 + 831	12. 988 + 578	13. 369 + 355	14. 796 + 109
15. 710 + 338	16. 140 + 584	17. 154 + 889	18. 865 + 464	19. 123 + 703	20. 794 + 929	21. 442 + 408
22. 402 + 875	23. 669 + 579	24. 860 + 654	25. 233 + 791	26. 563 + 555	27. 374 + 506	28. 614 + 245
29. 535 + 142	30. 754 + 637	31. 923 + 765	32. 986 + 125	33. 920 + 996	34. 614 + 651	35. 762 + 716
36. 168 + 104	37. 229 + 543	38. 395 + 866	39. 529 + 213	40. 908 + 790	41. 494 + 242	42. 874 + 918

Find the sum

Complete all the activities (Addition).

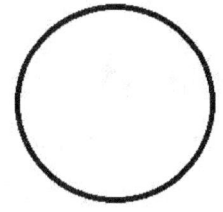

SCORE

| 1. | 976
 + 524
 1 500 | 2. | 620
 + 125
 745 | 3. | 892
 + 720
 1 612 | 4. | 774
 + 697
 1 471 | 5. | 363
 + 960
 1 323 | 6. | 581
 + 668
 1 249 | 7. | 540
 + 628
 1 168 |

8. 182   9. 707   10. 473   11. 276   12. 988   13. 369   14. 796
    + 191   + 923   + 737   + 831   + 578   + 355   + 109
    373   1 630   1 210   1 107   1 566   724   905

15. 710   16. 140   17. 154   18. 865   19. 123   20. 794   21. 442
     + 338   + 584   + 889   + 464   + 703   + 929   + 408
     1 048   724   1 043   1 329   826   1 723   850

22. 402   23. 669   24. 860   25. 233   26. 563   27. 374   28. 614
     + 875   + 579   + 654   + 791   + 555   + 506   + 245
     1 277   1 248   1 514   1 024   1 118   880   859

29. 535   30. 754   31. 923   32. 986   33. 920   34. 614   35. 762
     + 142   + 637   + 765   + 125   + 996   + 651   + 716
     677   1 391   1 688   1 111   1 916   1 265   1 478

36. 168   37. 229   38. 395   39. 529   40. 908   41. 494   42. 874
     + 104   + 543   + 866   + 213   + 790   + 242   + 918
     272   772   1 261   742   1 698   736   1 792

Name:............................ Date:........................

Find the sum

Complete all the activities (Addition).

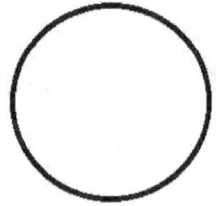

SCORE

1. 774 + 584	2. 310 + 647	3. 374 + 720	4. 928 + 980	5. 947 + 256	6. 106 + 676	7. 421 + 378
8. 955 + 672	9. 964 + 309	10. 891 + 516	11. 723 + 559	12. 426 + 830	13. 211 + 747	14. 714 + 422
15. 233 + 964	16. 243 + 781	17. 890 + 169	18. 350 + 436	19. 786 + 966	20. 376 + 142	21. 787 + 187
22. 156 + 972	23. 537 + 718	24. 546 + 252	25. 257 + 336	26. 825 + 285	27. 350 + 459	28. 295 + 315
29. 258 + 480	30. 536 + 627	31. 106 + 606	32. 847 + 983	33. 113 + 324	34. 786 + 352	35. 128 + 619
36. 885 + 997	37. 932 + 849	38. 728 + 636	39. 503 + 453	40. 393 + 492	41. 301 + 563	42. 265 + 896

Find the sum

Complete all the activities (Addition).

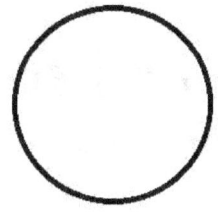

SCORE

1. 774 + 584 1 358	2. 310 + 647 957	3. 374 + 720 1 094	4. 928 + 980 1 908	5. 947 + 256 1 203	6. 106 + 676 782	7. 421 + 378 799
8. 955 + 672 1 627	9. 964 + 309 1 273	10. 891 + 516 1 407	11. 723 + 559 1 282	12. 426 + 830 1 256	13. 211 + 747 958	14. 714 + 422 1 136
15. 233 + 964 1 197	16. 243 + 781 1 024	17. 890 + 169 1 059	18. 350 + 436 786	19. 786 + 966 1 752	20. 376 + 142 518	21. 787 + 187 974
22. 156 + 972 1 128	23. 537 + 718 1 255	24. 546 + 252 798	25. 257 + 336 593	26. 825 + 285 1 110	27. 350 + 459 809	28. 295 + 315 610
29. 258 + 480 738	30. 536 + 627 1 163	31. 106 + 606 712	32. 847 + 983 1 830	33. 113 + 324 437	34. 786 + 352 1 138	35. 128 + 619 747
36. 885 + 997 1 882	37. 932 + 849 1 781	38. 728 + 636 1 364	39. 503 + 453 956	40. 393 + 492 885	41. 301 + 563 864	42. 265 + 896 1 161

Find the sum

Complete all the activities (Addition).

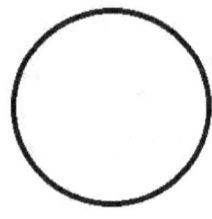

SCORE

1. 239 + 957	2. 532 + 531	3. 873 + 134	4. 278 + 406	5. 530 + 748	6. 136 + 281	7. 339 + 172
8. 959 + 313	9. 931 + 294	10. 224 + 637	11. 450 + 826	12. 352 + 500	13. 730 + 658	14. 812 + 804
15. 187 + 144	16. 426 + 542	17. 441 + 950	18. 843 + 155	19. 791 + 355	20. 478 + 993	21. 441 + 578
22. 568 + 284	23. 813 + 547	24. 396 + 430	25. 493 + 180	26. 560 + 107	27. 439 + 627	28. 546 + 412
29. 705 + 132	30. 800 + 315	31. 321 + 472	32. 319 + 486	33. 422 + 593	34. 838 + 388	35. 328 + 904
36. 520 + 676	37. 815 + 230	38. 296 + 871	39. 359 + 168	40. 686 + 284	41. 362 + 482	42. 411 + 107

Find the sum

Complete all the activities (Addition).

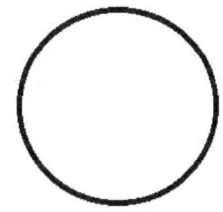

SCORE

| 1. | 239
+ 957
1 196 | 2. | 532
+ 531
1 063 | 3. | 873
+ 134
1 007 | 4. | 278
+ 406
684 | 5. | 530
+ 748
1 278 | 6. | 136
+ 281
417 | 7. | 339
+ 172
511 |

8.   959   9.   931   10. 224   11. 450   12. 352   13. 730   14. 812
    + 313   + 294   + 637   + 826   + 500   + 658   + 804
    1 272   1 225   861   1 276   852   1 388   1 616

15. 187   16. 426   17. 441   18. 843   19. 791   20. 478   21. 441
    + 144   + 542   + 950   + 155   + 355   + 993   + 578
     331   968   1 391   998   1 146   1 471   1 019

22. 568   23. 813   24. 396   25. 493   26. 560   27. 439   28. 546
    + 284   + 547   + 430   + 180   + 107   + 627   + 412
     852   1 360   826   673   667   1 066   958

29. 705   30. 800   31. 321   32. 319   33. 422   34. 838   35. 328
    + 132   + 315   + 472   + 486   + 593   + 388   + 904
     837   1 115   793   805   1 015   1 226   1 232

36. 520   37. 815   38. 296   39. 359   40. 686   41. 362   42. 411
    + 676   + 230   + 871   + 168   + 284   + 482   + 107
    1 196   1 045   1 167   527   970   844   518

Name:...................................   Date:...............................

Find the sum

Complete all the activities (Addition).

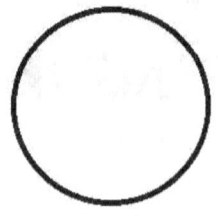

SCORE

1. 583 + 419	2. 203 + 814	3. 504 + 531	4. 448 + 760	5. 554 + 263	6. 771 + 968	7. 359 + 182
8. 860 + 320	9. 707 + 108	10. 202 + 589	11. 659 + 883	12. 291 + 309	13. 430 + 674	14. 153 + 638
15. 444 + 616	16. 952 + 839	17. 161 + 313	18. 902 + 103	19. 490 + 125	20. 883 + 796	21. 555 + 760
22. 222 + 652	23. 458 + 327	24. 612 + 276	25. 306 + 467	26. 575 + 354	27. 380 + 822	28. 447 + 699
29. 228 + 893	30. 483 + 829	31. 478 + 810	32. 416 + 937	33. 310 + 932	34. 907 + 146	35. 279 + 857
36. 829 + 683	37. 224 + 405	38. 133 + 898	39. 761 + 638	40. 851 + 203	41. 161 + 876	42. 431 + 977

Name:................................ Date:................................

Find the sum

Complete all the activities (Addition).

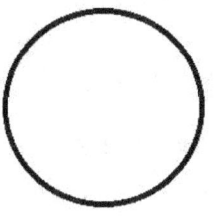

SCORE

| 1. 583
+ 419
1 002 | 2. 203
+ 814
1 017 | 3. 504
+ 531
1 035 | 4. 448
+ 760
1 208 | 5. 554
+ 263
817 | 6. 771
+ 968
1 739 | 7. 359
+ 182
541 |

| 8. 860
+ 320
1 180 | 9. 707
+ 108
815 | 10. 202
+ 589
791 | 11. 659
+ 883
1 542 | 12. 291
+ 309
600 | 13. 430
+ 674
1 104 | 14. 153
+ 638
791 |

| 15. 444
+ 616
1 060 | 16. 952
+ 839
1 791 | 17. 161
+ 313
474 | 18. 902
+ 103
1 005 | 19. 490
+ 125
615 | 20. 883
+ 796
1 679 | 21. 555
+ 760
1 315 |

| 22. 222
+ 652
874 | 23. 458
+ 327
785 | 24. 612
+ 276
888 | 25. 306
+ 467
773 | 26. 575
+ 354
929 | 27. 380
+ 822
1 202 | 28. 447
+ 699
1 146 |

| 29. 228
+ 893
1 121 | 30. 483
+ 829
1 312 | 31. 478
+ 810
1 288 | 32. 416
+ 937
1 353 | 33. 310
+ 932
1 242 | 34. 907
+ 146
1 053 | 35. 279
+ 857
1 136 |

| 36. 829
+ 683
1 512 | 37. 224
+ 405
629 | 38. 133
+ 898
1 031 | 39. 761
+ 638
1 399 | 40. 851
+ 203
1 054 | 41. 161
+ 876
1 037 | 42. 431
+ 977
1 408 |

Find the sum

Complete all the activities (Addition).

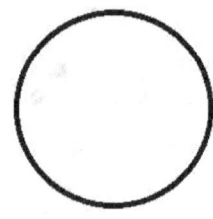

SCORE

1. 141 + 699	2. 559 + 338	3. 865 + 303	4. 874 + 597	5. 640 + 154	6. 418 + 744	7. 444 + 284
8. 442 + 525	9. 419 + 566	10. 996 + 448	11. 637 + 892	12. 230 + 974	13. 788 + 548	14. 937 + 589
15. 142 + 380	16. 768 + 601	17. 479 + 438	18. 896 + 611	19. 472 + 669	20. 280 + 975	21. 982 + 332
22. 143 + 170	23. 985 + 388	24. 554 + 577	25. 296 + 426	26. 784 + 594	27. 879 + 982	28. 440 + 248
29. 569 + 584	30. 549 + 417	31. 365 + 385	32. 555 + 360	33. 385 + 817	34. 365 + 106	35. 387 + 830
36. 153 + 749	37. 386 + 121	38. 466 + 405	39. 849 + 160	40. 987 + 500	41. 993 + 460	42. 393 + 715

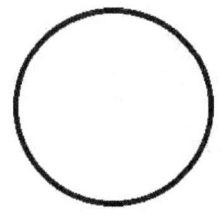

Find the sum

Complete all the activities (Addition).

| 1. | 141
+ 699
840 | 2. | 559
+ 338
897 | 3. | 865
+ 303
1 168 | 4. | 874
+ 597
1 471 | 5. | 640
+ 154
794 | 6. | 418
+ 744
1 162 | 7. | 444
+ 284
728 |

| 8. | 442
+ 525
967 | 9. | 419
+ 566
985 | 10. | 996
+ 448
1 444 | 11. | 637
+ 892
1 529 | 12. | 230
+ 974
1 204 | 13. | 788
+ 548
1 336 | 14. | 937
+ 589
1 526 |

| 15. | 142
+ 380
522 | 16. | 768
+ 601
1 369 | 17. | 479
+ 438
917 | 18. | 896
+ 611
1 507 | 19. | 472
+ 669
1 141 | 20. | 280
+ 975
1 255 | 21. | 982
+ 332
1 314 |

| 22. | 143
+ 170
313 | 23. | 985
+ 388
1 373 | 24. | 554
+ 577
1 131 | 25. | 296
+ 426
722 | 26. | 784
+ 594
1 378 | 27. | 879
+ 982
1 861 | 28. | 440
+ 248
688 |

| 29. | 569
+ 584
1 153 | 30. | 549
+ 417
966 | 31. | 365
+ 385
750 | 32. | 555
+ 360
915 | 33. | 385
+ 817
1 202 | 34. | 365
+ 106
471 | 35. | 387
+ 830
1 217 |

| 36. | 153
+ 749
902 | 37. | 386
+ 121
507 | 38. | 466
+ 405
871 | 39. | 849
+ 160
1 009 | 40. | 987
+ 500
1 487 | 41. | 993
+ 460
1 453 | 42. | 393
+ 715
1 108 |

Find the sum

Complete all the activities (Addition).

Name:......................................   Date:...............................

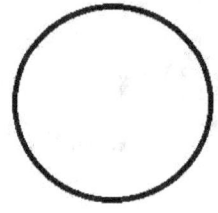

SCORE

| 1. | 596
 + 274 | 2. | 721
 + 496 | 3. | 997
 + 748 | 4. | 142
 + 887 | 5. | 220
 + 862 | 6. | 710
 + 696 | 7. | 776
 + 781 |

1. 596
    + 274

2. 721
    + 496

3. 997
    + 748

4. 142
    + 887

5. 220
    + 862

6. 710
    + 696

7. 776
    + 781

8. 784
    + 269

9. 123
    + 547

10. 290
    + 465

11. 373
    + 135

12. 308
    + 857

13. 616
    + 135

14. 398
    + 509

15. 596
    + 149

16. 423
    + 904

17. 992
    + 839

18. 764
    + 211

19. 196
    + 868

20. 108
    + 637

21. 788
    + 265

22. 629
    + 524

23. 411
    + 768

24. 628
    + 168

25. 375
    + 819

26. 478
    + 546

27. 270
    + 927

28. 442
    + 561

29. 965
    + 863

30. 879
    + 963

31. 450
    + 632

32. 391
    + 171

33. 539
    + 804

34. 704
    + 936

35. 118
    + 673

36. 173
    + 253

37. 539
    + 652

38. 676
    + 208

39. 505
    + 783

40. 547
    + 847

41. 506
    + 330

42. 234
    + 776

Find the sum

Complete all the activities (Addition).

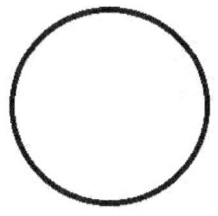

SCORE

1. 596 + 274 870	2. 721 + 496 1 217	3. 997 + 748 1 745	4. 142 + 887 1 029	5. 220 + 862 1 082	6. 710 + 696 1 406	7. 776 + 781 1 557
8. 784 + 269 1 053	9. 123 + 547 670	10. 290 + 465 755	11. 373 + 135 508	12. 308 + 857 1 165	13. 616 + 135 751	14. 398 + 509 907
15. 596 + 149 745	16. 423 + 904 1 327	17. 992 + 839 1 831	18. 764 + 211 975	19. 196 + 868 1 064	20. 108 + 637 745	21. 788 + 265 1 053
22. 629 + 524 1 153	23. 411 + 768 1 179	24. 628 + 168 796	25. 375 + 819 1 194	26. 478 + 546 1 024	27. 270 + 927 1 197	28. 442 + 561 1 003
29. 965 + 863 1 828	30. 879 + 963 1 842	31. 450 + 632 1 082	32. 391 + 171 562	33. 539 + 804 1 343	34. 704 + 936 1 640	35. 118 + 673 791
36. 173 + 253 426	37. 539 + 652 1 191	38. 676 + 208 884	39. 505 + 783 1 288	40. 547 + 847 1 394	41. 506 + 330 836	42. 234 + 776 1 010

Name:.. Date:...............................

Find the sum

Complete all the activities (Addition).

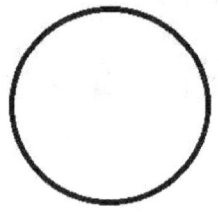

SCORE

1. 703
    + 316
    ----

2. 580
    + 663
    ----

3. 973
    + 823
    ----

4. 920
    + 197
    ----

5. 923
    + 476
    ----

6. 687
    + 837
    ----

7. 244
    + 531
    ----

8. 718
    + 308
    ----

9. 493
    + 158
    ----

10. 343
    + 142
    ----

11. 772
    + 211
    ----

12. 651
    + 748
    ----

13. 683
    + 994
    ----

14. 396
    + 227
    ----

15. 462
    + 587
    ----

16. 243
    + 736
    ----

17. 790
    + 619
    ----

18. 734
    + 163
    ----

19. 219
    + 580
    ----

20. 327
    + 509
    ----

21. 544
    + 470
    ----

22. 836
    + 310
    ----

23. 553
    + 137
    ----

24. 673
    + 800
    ----

25. 229
    + 994
    ----

26. 421
    + 217
    ----

27. 539
    + 888
    ----

28. 469
    + 592
    ----

29. 664
    + 832
    ----

30. 979
    + 978
    ----

31. 527
    + 793
    ----

32. 733
    + 436
    ----

33. 333
    + 950
    ----

34. 147
    + 624
    ----

35. 796
    + 654
    ----

36. 189
    + 912
    ----

37. 831
    + 833
    ----

38. 596
    + 402
    ----

39. 374
    + 441
    ----

40. 294
    + 646
    ----

41. 421
    + 595
    ----

42. 997
    + 467

Name:................................ Date:................................

Find the sum

Complete all the activities (Addition).

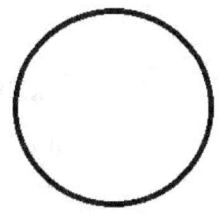

SCORE

1. 703 + 316 1 019	2. 580 + 663 1 243	3. 973 + 823 1 796	4. 920 + 197 1 117	5. 923 + 476 1 399	6. 687 + 837 1 524	7. 244 + 531 775
8. 718 + 308 1 026	9. 493 + 158 651	10. 343 + 142 485	11. 772 + 211 983	12. 651 + 748 1 399	13. 683 + 994 1 677	14. 396 + 227 623
15. 462 + 587 1 049	16. 243 + 736 979	17. 790 + 619 1 409	18. 734 + 163 897	19. 219 + 580 799	20. 327 + 509 836	21. 544 + 470 1 014
22. 836 + 310 1 146	23. 553 + 137 690	24. 673 + 800 1 473	25. 229 + 994 1 223	26. 421 + 217 638	27. 539 + 888 1 427	28. 469 + 592 1 061
29. 664 + 832 1 496	30. 979 + 978 1 957	31. 527 + 793 1 320	32. 733 + 436 1 169	33. 333 + 950 1 283	34. 147 + 624 771	35. 796 + 654 1 450
36. 189 + 912 1 101	37. 831 + 833 1 664	38. 596 + 402 998	39. 374 + 441 815	40. 294 + 646 940	41. 421 + 595 1 016	42. 997 + 467 1 464

Find the sum

Complete all the activities (Addition).

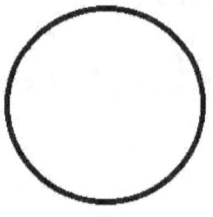

SCORE

1. 266 + 125	2. 297 + 901	3. 250 + 242	4. 499 + 386	5. 600 + 347	6. 981 + 750	7. 736 + 500
8. 976 + 602	9. 989 + 149	10. 104 + 574	11. 905 + 508	12. 386 + 704	13. 392 + 167	14. 318 + 323
15. 637 + 433	16. 657 + 931	17. 795 + 366	18. 270 + 308	19. 753 + 245	20. 474 + 654	21. 256 + 801
22. 413 + 583	23. 851 + 845	24. 665 + 873	25. 464 + 691	26. 656 + 372	27. 591 + 523	28. 561 + 757
29. 739 + 592	30. 609 + 183	31. 923 + 276	32. 748 + 267	33. 689 + 150	34. 953 + 469	35. 330 + 769
36. 848 + 844	37. 674 + 449	38. 418 + 204	39. 195 + 801	40. 852 + 546	41. 501 + 444	42. 858 + 945

Find the sum

Complete all the activities (Addition).

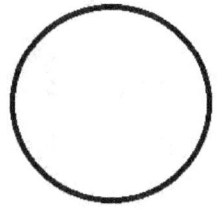

SCORE

1. 266 + 125 391	2. 297 + 901 1 198	3. 250 + 242 492	4. 499 + 386 885	5. 600 + 347 947	6. 981 + 750 1 731	7. 736 + 500 1 236
8. 976 + 602 1 578	9. 989 + 149 1 138	10. 104 + 574 678	11. 905 + 508 1 413	12. 386 + 704 1 090	13. 392 + 167 559	14. 318 + 323 641
15. 637 + 433 1 070	16. 657 + 931 1 588	17. 795 + 366 1 161	18. 270 + 308 578	19. 753 + 245 998	20. 474 + 654 1 128	21. 256 + 801 1 057
22. 413 + 583 996	23. 851 + 845 1 696	24. 665 + 873 1 538	25. 464 + 691 1 155	26. 656 + 372 1 028	27. 591 + 523 1 114	28. 561 + 757 1 318
29. 739 + 592 1 331	30. 609 + 183 792	31. 923 + 276 1 199	32. 748 + 267 1 015	33. 689 + 150 839	34. 953 + 469 1 422	35. 330 + 769 1 099
36. 848 + 844 1 692	37. 674 + 449 1 123	38. 418 + 204 622	39. 195 + 801 996	40. 852 + 546 1 398	41. 501 + 444 945	42. 858 + 945 1 803

Name:.................................   Date:....................................

Find the sum

Complete all the activities (Addition).

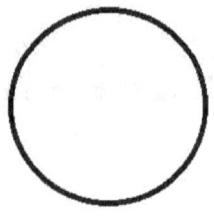

SCORE

1. 776 + 385	2. 440 + 372	3. 804 + 512	4. 568 + 262	5. 636 + 391	6. 729 + 979	7. 153 + 245
8. 893 + 802	9. 891 + 296	10. 643 + 694	11. 843 + 811	12. 697 + 242	13. 205 + 457	14. 661 + 937
15. 739 + 163	16. 681 + 580	17. 417 + 803	18. 490 + 404	19. 709 + 681	20. 664 + 817	21. 301 + 686
22. 308 + 382	23. 718 + 438	24. 150 + 949	25. 907 + 150	26. 654 + 306	27. 800 + 264	28. 202 + 964
29. 687 + 297	30. 267 + 447	31. 850 + 630	32. 522 + 573	33. 710 + 250	34. 662 + 349	35. 754 + 589
36. 193 + 310	37. 630 + 825	38. 493 + 257	39. 603 + 415	40. 523 + 766	41. 407 + 712	42. 129 + 689

Find the sum

Complete all the activities (Addition).

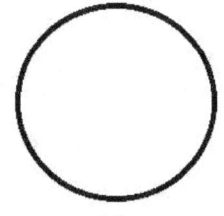

SCORE

1. 776 + 385 1 161	2. 440 + 372 812	3. 804 + 512 1 316	4. 568 + 262 830	5. 636 + 391 1 027	6. 729 + 979 1 708	7. 153 + 245 398
8. 893 + 802 1 695	9. 891 + 296 1 187	10. 643 + 694 1 337	11. 843 + 811 1 654	12. 697 + 242 939	13. 205 + 457 662	14. 661 + 937 1 598
15. 739 + 163 902	16. 681 + 580 1 261	17. 417 + 803 1 220	18. 490 + 404 894	19. 709 + 681 1 390	20. 664 + 817 1 481	21. 301 + 686 987
22. 308 + 382 690	23. 718 + 438 1 156	24. 150 + 949 1 099	25. 907 + 150 1 057	26. 654 + 306 960	27. 800 + 264 1 064	28. 202 + 964 1 166
29. 687 + 297 984	30. 267 + 447 714	31. 850 + 630 1 480	32. 522 + 573 1 095	33. 710 + 250 960	34. 662 + 349 1 011	35. 754 + 589 1 343
36. 193 + 310 503	37. 630 + 825 1 455	38. 493 + 257 750	39. 603 + 415 1 018	40. 523 + 766 1 289	41. 407 + 712 1 119	42. 129 + 689 818

Name:................................   Date:.........................

Find the sum

Complete all the activities (Addition).

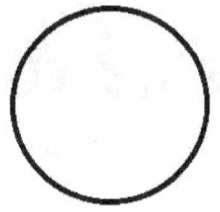

SCORE

1. 508 + 475	2. 778 + 253	3. 701 + 526	4. 716 + 511	5. 515 + 318	6. 966 + 669	7. 265 + 735
8. 671 + 569	9. 775 + 800	10. 295 + 418	11. 956 + 444	12. 674 + 974	13. 770 + 944	14. 461 + 372
15. 773 + 499	16. 140 + 238	17. 601 + 832	18. 165 + 335	19. 470 + 510	20. 648 + 328	21. 873 + 680
22. 794 + 413	23. 858 + 823	24. 991 + 867	25. 743 + 953	26. 991 + 606	27. 517 + 711	28. 732 + 164
29. 800 + 803	30. 482 + 756	31. 165 + 444	32. 737 + 115	33. 278 + 827	34. 168 + 432	35. 670 + 437
36. 198 + 365	37. 252 + 966	38. 544 + 376	39. 649 + 362	40. 215 + 946	41. 952 + 593	42. 628 + 819

Name:................................ Date:................................

Find the sum

Complete all the activities (Addition).

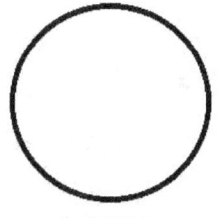

SCORE

1.	508	2.	778	3.	701	4.	716	5.	515	6.	966	7.	265
	+ 475		+ 253		+ 526		+ 511		+ 318		+ 669		+ 735
	983		1 031		1 227		1 227		833		1 635		1 000

8.	671	9.	775	10.	295	11.	956	12.	674	13.	770	14.	461
	+ 569		+ 800		+ 418		+ 444		+ 974		+ 944		+ 372
	1 240		1 575		713		1 400		1 648		1 714		833

15.	773	16.	140	17.	601	18.	165	19.	470	20.	648	21.	873
	+ 499		+ 238		+ 832		+ 335		+ 510		+ 328		+ 680
	1 272		378		1 433		500		980		976		1 553

22.	794	23.	858	24.	991	25.	743	26.	991	27.	517	28.	732
	+ 413		+ 823		+ 867		+ 953		+ 606		+ 711		+ 164
	1 207		1 681		1 858		1 696		1 597		1 228		896

29.	800	30.	482	31.	165	32.	737	33.	278	34.	168	35.	670
	+ 803		+ 756		+ 444		+ 115		+ 827		+ 432		+ 437
	1 603		1 238		609		852		1 105		600		1 107

36.	198	37.	252	38.	544	39.	649	40.	215	41.	952	42.	628
	+ 365		+ 966		+ 376		+ 362		+ 946		+ 593		+ 819
	563		1 218		920		1 011		1 161		1 545		1 447

Name:.................................   Date:.................................

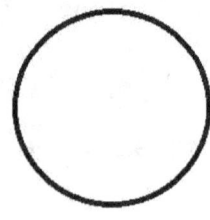

SCORE

Find the sum

Complete all the activities (Addition).

1.	2.	3.	4.	5.	6.	7.
508 + 533	721 + 889	200 + 101	534 + 666	128 + 891	264 + 529	942 + 779

8.	9.	10.	11.	12.	13.	14.
235 + 131	754 + 659	297 + 406	855 + 725	454 + 550	267 + 456	514 + 538

15.	16.	17.	18.	19.	20.	21.
688 + 535	802 + 570	815 + 917	607 + 744	207 + 278	331 + 670	902 + 531

22.	23.	24.	25.	26.	27.	28.
337 + 360	802 + 115	432 + 869	227 + 184	292 + 393	397 + 321	963 + 781

29.	30.	31.	32.	33.	34.	35.
179 + 271	723 + 743	586 + 995	977 + 592	189 + 692	712 + 125	272 + 490

36.	37.	38.	39.	40.	41.	42.
924 + 576	428 + 626	504 + 424	488 + 193	747 + 251	124 + 644	412 + 604

Find the sum

Complete all the activities (Addition).

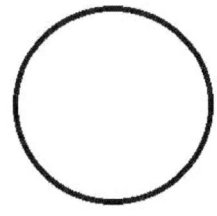

SCORE

1. 508 + 533 1 041	2. 721 + 889 1 610	3. 200 + 101 301	4. 534 + 666 1 200	5. 128 + 891 1 019	6. 264 + 529 793	7. 942 + 779 1 721
8. 235 + 131 366	9. 754 + 659 1 413	10. 297 + 406 703	11. 855 + 725 1 580	12. 454 + 550 1 004	13. 267 + 456 723	14. 514 + 538 1 052
15. 688 + 535 1 223	16. 802 + 570 1 372	17. 815 + 917 1 732	18. 607 + 744 1 351	19. 207 + 278 485	20. 331 + 670 1 001	21. 902 + 531 1 433
22. 337 + 360 697	23. 802 + 115 917	24. 432 + 869 1 301	25. 227 + 184 411	26. 292 + 393 685	27. 397 + 321 718	28. 963 + 781 1 744
29. 179 + 271 450	30. 723 + 743 1 466	31. 586 + 995 1 581	32. 977 + 592 1 569	33. 189 + 692 881	34. 712 + 125 837	35. 272 + 490 762
36. 924 + 576 1 500	37. 428 + 626 1 054	38. 504 + 424 928	39. 488 + 193 681	40. 747 + 251 998	41. 124 + 644 768	42. 412 + 604 1 016

Section 2

Find the difference

Complete all the activities (Subtraction)

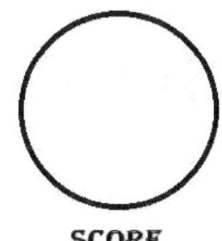

SCORE

| 1. 622
- 388
------- | 2. 282
- 185
------- | 3. 359
- 294
------- | 4. 681
- 419
------- | 5. 497
- 263
------- | 6. 123
- 109
------- | 7. 867
- 395
------- | 8. 815
- 491
------- |

| 9. 194
- 174
------- | 10. 407
- 335
------- | 11. 406
- 270
------- | 12. 308
- 160
------- | 13. 377
- 213
------- | 14. 921
- 367
------- | 15. 105
- 101
------- | 16. 505
- 163
------- |

| 17. 318
- 137
------- | 18. 329
- 102
------- | 19. 155
- 127
------- | 20. 450
- 421
------- | 21. 456
- 373
------- | 22. 199
- 186
------- | 23. 677
- 385
------- | 24. 769
- 451
------- |

| 25. 105
- 103
------- | 26. 482
- 453
------- | 27. 883
- 800
------- | 28. 526
- 111
------- | 29. 389
- 364
------- | 30. 241
- 224
------- | 31. 851
- 341
------- | 32. 588
- 240
------- |

| 33. 964
- 852
------- | 34. 231
- 190
------- | 35. 217
- 192
------- | 36. 536
- 367
------- | 37. 455
- 250
------- | 38. 433
- 291
------- | 39. 501
- 245
------- | 40. 680
- 573
------- |

| 41. 697
- 332
------- | 42. 387
- 188
------- | 43. 756
- 446
------- | 44. 356
- 189
------- | 45. 725
- 243
------- | 46. 719
- 191
------- | 47. 402
- 254
------- | 48. 174
- 106
------- |

Find the difference

Complete all the activities (Subtraction)

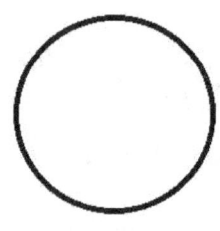

SCORE

1. 622 − 388 = 234	2. 282 − 185 = 97	3. 359 − 294 = 65	4. 681 − 419 = 262	5. 497 − 263 = 234	6. 123 − 109 = 14	7. 867 − 395 = 472	8. 815 − 491 = 324
9. 194 − 174 = 20	10. 407 − 335 = 72	11. 406 − 270 = 136	12. 308 − 160 = 148	13. 377 − 213 = 164	14. 921 − 367 = 554	15. 105 − 101 = 4	16. 505 − 163 = 342
17. 318 − 137 = 181	18. 329 − 102 = 227	19. 155 − 127 = 28	20. 450 − 421 = 29	21. 456 − 373 = 83	22. 199 − 186 = 13	23. 677 − 385 = 292	24. 769 − 451 = 318
25. 105 − 103 = 2	26. 482 − 453 = 29	27. 883 − 800 = 83	28. 526 − 111 = 415	29. 389 − 364 = 25	30. 241 − 224 = 17	31. 851 − 341 = 510	32. 588 − 240 = 348
33. 964 − 852 = 112	34. 231 − 190 = 41	35. 217 − 192 = 25	36. 536 − 367 = 169	37. 455 − 250 = 205	38. 433 − 291 = 142	39. 501 − 245 = 256	40. 680 − 573 = 107
41. 697 − 332 = 365	42. 387 − 188 = 199	43. 756 − 446 = 310	44. 356 − 189 = 167	45. 725 − 243 = 482	46. 719 − 191 = 528	47. 402 − 254 = 148	48. 174 − 106 = 68

Name:................................   Date:................................

Find the difference

Complete all the activities (Subtraction)

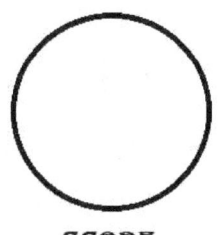

SCORE

1. 902 - 378	2. 355 - 278	3. 728 - 209	4. 530 - 497	5. 755 - 316	6. 568 - 198	7. 618 - 255	8. 632 - 115
--------	--------	--------	--------	--------	--------	--------	--------
9. 499 - 473	10. 114 - 112	11. 129 - 126	12. 410 - 115	13. 967 - 388	14. 963 - 824	15. 773 - 373	16. 201 - 144
--------	--------	--------	--------	--------	--------	--------	--------
17. 385 - 173	18. 418 - 410	19. 169 - 131	20. 479 - 315	21. 943 - 655	22. 763 - 668	23. 655 - 267	24. 640 - 521
--------	--------	--------	--------	--------	--------	--------	--------
25. 857 - 487	26. 970 - 149	27. 895 - 839	28. 972 - 965	29. 102 - 100	30. 759 - 341	31. 292 - 223	32. 257 - 151
--------	--------	--------	--------	--------	--------	--------	--------
33. 295 - 117	34. 903 - 356	35. 950 - 626	36. 956 - 953	37. 561 - 307	38. 916 - 405	39. 246 - 171	40. 488 - 357
--------	--------	--------	--------	--------	--------	--------	--------
41. 757 - 218	42. 996 - 313	43. 495 - 122	44. 593 - 291	45. 357 - 193	46. 974 - 836	47. 742 - 136	48. 417 - 194
--------	--------	--------	--------	--------	--------	--------	--------

Name:.................................. Date:..............................

Find the difference

Complete all the activities (Subtraction)

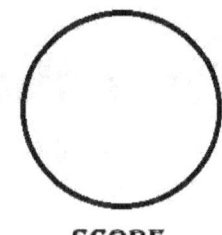

SCORE

1. 902 − 378 = 524	2. 355 − 278 = 77	3. 728 − 209 = 519	4. 530 − 497 = 33	5. 755 − 316 = 439	6. 568 − 198 = 370	7. 618 − 255 = 363	8. 632 − 115 = 517
9. 499 − 473 = 26	10. 114 − 112 = 2	11. 129 − 126 = 3	12. 410 − 115 = 295	13. 967 − 388 = 579	14. 963 − 824 = 139	15. 773 − 373 = 400	16. 201 − 144 = 57
17. 385 − 173 = 212	18. 418 − 410 = 8	19. 169 − 131 = 38	20. 479 − 315 = 164	21. 943 − 655 = 288	22. 763 − 668 = 95	23. 655 − 267 = 388	24. 640 − 521 = 119
25. 857 − 487 = 370	26. 970 − 149 = 821	27. 895 − 839 = 56	28. 972 − 965 = 7	29. 102 − 100 = 2	30. 759 − 341 = 418	31. 292 − 223 = 69	32. 257 − 151 = 106
33. 295 − 117 = 178	34. 903 − 356 = 547	35. 950 − 626 = 324	36. 956 − 953 = 3	37. 561 − 307 = 254	38. 916 − 405 = 511	39. 246 − 171 = 75	40. 488 − 357 = 131
41. 757 − 218 = 539	42. 996 − 313 = 683	43. 495 − 122 = 373	44. 593 − 291 = 302	45. 357 − 193 = 164	46. 974 − 836 = 138	47. 742 − 136 = 606	48. 417 − 194 = 223

Name:..................................   Date:....................................

Find the difference

Complete all the activities (Subtraction)

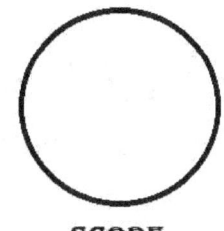

SCORE

1.   535
  - 281
-------- 	
	2. 675
	  - 404
	-------- 	
		3. 467
		  - 203
		-------- 	
			4. 849
			  - 268
			-------- 	
				5. 603
				  - 588
				-------- 	
					6. 589
					  - 313
					-------- 	
						7. 120
						  - 112
						-------- 	
							8. 846
							  - 483
							-------- 	

9.   822
  - 151
-------- 	
	10. 671
	  - 119
	-------- 	
		11. 679
		  - 642
		-------- 	
			12. 894
			  - 227
			-------- 	
				13. 586
				  - 346
				-------- 	
					14. 130
					  - 118
					-------- 	
						15. 681
						  - 164
						-------- 	
							16. 354
							  - 216
							-------- 	

17.   187
  - 153
-------- 	
	18. 901
	  - 698
	-------- 	
		19. 863
		  - 790
		-------- 	
			20. 329
			  - 203
			-------- 	
				21. 412
				  - 341
				-------- 	
					22. 710
					  - 386
					-------- 	
						23. 560
						  - 152
						-------- 	
							24. 428
							  - 326
							-------- 	

25.   578
  - 319
-------- 	
	26. 136
	  - 123
	-------- 	
		27. 478
		  - 221
		-------- 	
			28. 688
			  - 421
			-------- 	
				29. 579
				  - 330
				-------- 	
					30. 988
					  - 901
					-------- 	
						31. 323
						  - 156
						-------- 	
							32. 619
							  - 334
							-------- 	

33.   719
  - 154
-------- 	
	34. 404
	  - 225
	-------- 	
		35. 734
		  - 153
		-------- 	
			36. 697
			  - 637
			-------- 	
				37. 833
				  - 236
				-------- 	
					38. 717
					  - 638
					-------- 	
						39. 333
						  - 197
						-------- 	
							40. 683
							  - 349
							-------- 	

41.   227
  - 144
-------- 	
	42. 148
	  - 109
	-------- 	
		43. 163
		  - 140
		-------- 	
			44. 133
			  - 126
			-------- 	
				45. 959
				  - 396
				-------- 	
					46. 293
					  - 171
					-------- 	
						47. 924
						  - 390
						-------- 	
							48. 281
							  - 118
							-------- 	

Name:.................................   Date:...............................

Find the difference

Complete all the activities (Subtraction)

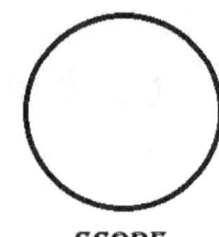

SCORE

1. 535 − 281 = 254	2. 675 − 404 = 271	3. 467 − 203 = 264	4. 849 − 268 = 581	5. 603 − 588 = 15	6. 589 − 313 = 276	7. 120 − 112 = 8	8. 846 − 483 = 363
9. 822 − 151 = 671	10. 671 − 119 = 552	11. 679 − 642 = 37	12. 894 − 227 = 667	13. 586 − 346 = 240	14. 130 − 118 = 12	15. 681 − 164 = 517	16. 354 − 216 = 138
17. 187 − 153 = 34	18. 901 − 698 = 203	19. 863 − 790 = 73	20. 329 − 203 = 126	21. 412 − 341 = 71	22. 710 − 386 = 324	23. 560 − 152 = 408	24. 428 − 326 = 102
25. 578 − 319 = 259	26. 136 − 123 = 13	27. 478 − 221 = 257	28. 688 − 421 = 267	29. 579 − 330 = 249	30. 988 − 901 = 87	31. 323 − 156 = 167	32. 619 − 334 = 285
33. 719 − 154 = 565	34. 404 − 225 = 179	35. 734 − 153 = 581	36. 697 − 637 = 60	37. 833 − 236 = 597	38. 717 − 638 = 79	39. 333 − 197 = 136	40. 683 − 349 = 334
41. 227 − 144 = 83	42. 148 − 109 = 39	43. 163 − 140 = 23	44. 133 − 126 = 7	45. 959 − 396 = 563	46. 293 − 171 = 122	47. 924 − 390 = 534	48. 281 − 118 = 163

Find the difference

Complete all the activities (Subtraction)

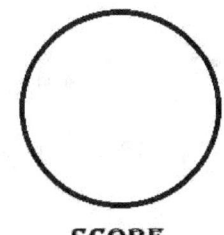

SCORE

1. 275 - 140 --------	2. 801 - 398 --------	3. 796 - 763 --------	4. 985 - 496 --------	5. 814 - 201 --------	6. 845 - 275 --------	7. 669 - 545 --------	8. 359 - 130 --------
9. 847 - 519 --------	10. 784 - 719 --------	11. 277 - 169 --------	12. 442 - 117 --------	13. 145 - 125 --------	14. 510 - 248 --------	15. 881 - 330 --------	16. 983 - 509 --------
17. 985 - 621 --------	18. 708 - 119 --------	19. 671 - 228 --------	20. 687 - 211 --------	21. 422 - 295 --------	22. 454 - 429 --------	23. 827 - 683 --------	24. 229 - 122 --------
25. 151 - 105 --------	26. 584 - 272 --------	27. 631 - 257 --------	28. 910 - 753 --------	29. 862 - 655 --------	30. 437 - 172 --------	31. 545 - 377 --------	32. 620 - 360 --------
33. 242 - 152 --------	34. 326 - 110 --------	35. 196 - 125 --------	36. 533 - 363 --------	37. 916 - 566 --------	38. 592 - 391 --------	39. 995 - 504 --------	40. 317 - 154 --------
41. 476 - 207 --------	42. 374 - 312 --------	43. 780 - 526 --------	44. 884 - 639 --------	45. 718 - 638 --------	46. 265 - 250 --------	47. 863 - 436 --------	48. 724 - 559 --------

Name:................................ Date:................................

Find the difference

Complete all the activities (Subtraction)

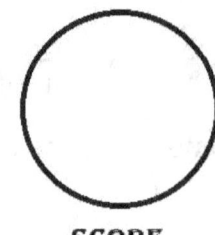

SCORE

1. 275 - 140 135	2. 801 - 398 403	3. 796 - 763 33	4. 985 - 496 489	5. 814 - 201 613	6. 845 - 275 570	7. 669 - 545 124	8. 359 - 130 229
9. 847 - 519 328	10. 784 - 719 65	11. 277 - 169 108	12. 442 - 117 325	13. 145 - 125 20	14. 510 - 248 262	15. 881 - 330 551	16. 983 - 509 474
17. 985 - 621 364	18. 708 - 119 589	19. 671 - 228 443	20. 687 - 211 476	21. 422 - 295 127	22. 454 - 429 25	23. 827 - 683 144	24. 229 - 122 107
25. 151 - 105 46	26. 584 - 272 312	27. 631 - 257 374	28. 910 - 753 157	29. 862 - 655 207	30. 437 - 172 265	31. 545 - 377 168	32. 620 - 360 260
33. 242 - 152 90	34. 326 - 110 216	35. 196 - 125 71	36. 533 - 363 170	37. 916 - 566 350	38. 592 - 391 201	39. 995 - 504 491	40. 317 - 154 163
41. 476 - 207 269	42. 374 - 312 62	43. 780 - 526 254	44. 884 - 639 245	45. 718 - 638 80	46. 265 - 250 15	47. 863 - 436 427	48. 724 - 559 165

Name:................................   Date:.................................

Find the difference

Complete all the activities (Subtraction)

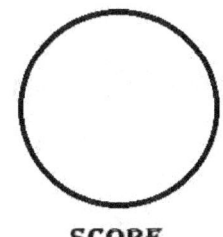

SCORE

| 1. | 124
- 121
-------- | 2. | 793
- 407
-------- | 3. | 595
- 149
-------- | 4. | 160
- 148
-------- | 5. | 653
- 615
-------- | 6. | 503
- 108
-------- | 7. | 774
- 309
-------- | 8. | 829
- 389
-------- |

| 9. | 538
- 155
-------- | 10. | 814
- 251
-------- | 11. | 386
- 331
-------- | 12. | 768
- 455
-------- | 13. | 543
- 349
-------- | 14. | 420
- 176
-------- | 15. | 166
- 138
-------- | 16. | 781
- 419
-------- |

| 17. | 888
- 740
-------- | 18. | 608
- 151
-------- | 19. | 711
- 587
-------- | 20. | 728
- 300
-------- | 21. | 295
- 199
-------- | 22. | 241
- 216
-------- | 23. | 260
- 236
-------- | 24. | 944
- 419
-------- |

| 25. | 154
- 142
-------- | 26. | 477
- 373
-------- | 27. | 680
- 380
-------- | 28. | 475
- 183
-------- | 29. | 237
- 133
-------- | 30. | 954
- 943
-------- | 31. | 768
- 654
-------- | 32. | 752
- 126
-------- |

| 33. | 256
- 253
-------- | 34. | 228
- 220
-------- | 35. | 770
- 576
-------- | 36. | 745
- 545
-------- | 37. | 484
- 283
-------- | 38. | 994
- 826
-------- | 39. | 683
- 635
-------- | 40. | 417
- 303
-------- |

| 41. | 662
- 650
-------- | 42. | 259
- 146
-------- | 43. | 241
- 105
-------- | 44. | 839
- 392
-------- | 45. | 455
- 222
-------- | 46. | 953
- 476
-------- | 47. | 862
- 641
-------- | 48. | 617
- 300
-------- |

Name:................................ Date:...................................

Find the difference

Complete all the activities (Subtraction)

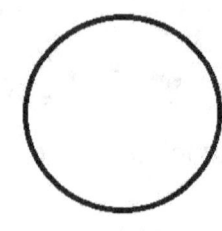

SCORE

1. 124 - 121 3	2. 793 - 407 386	3. 595 - 149 446	4. 160 - 148 12	5. 653 - 615 38	6. 503 - 108 395	7. 774 - 309 465	8. 829 - 389 440
9. 538 - 155 383	10. 814 - 251 563	11. 386 - 331 55	12. 768 - 455 313	13. 543 - 349 194	14. 420 - 176 244	15. 166 - 138 28	16. 781 - 419 362
17. 888 - 740 148	18. 608 - 151 457	19. 711 - 587 124	20. 728 - 300 428	21. 295 - 199 96	22. 241 - 216 25	23. 260 - 236 24	24. 944 - 419 525
25. 154 - 142 12	26. 477 - 373 104	27. 680 - 380 300	28. 475 - 183 292	29. 237 - 133 104	30. 954 - 943 11	31. 768 - 654 114	32. 752 - 126 626
33. 256 - 253 3	34. 228 - 220 8	35. 770 - 576 194	36. 745 - 545 200	37. 484 - 283 201	38. 994 - 826 168	39. 683 - 635 48	40. 417 - 303 114
41. 662 - 650 12	42. 259 - 146 113	43. 241 - 105 136	44. 839 - 392 447	45. 455 - 222 233	46. 953 - 476 477	47. 862 - 641 221	48. 617 - 300 317

Find the difference

Complete all the activities (Subtraction)

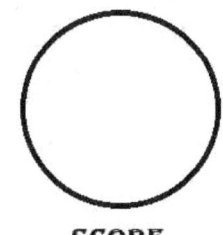

SCORE

1. 801 - 183 --------	2. 685 - 216 --------	3. 738 - 100 --------	4. 669 - 299 --------	5. 162 - 145 --------	6. 366 - 145 --------	7. 538 - 365 --------	8. 613 - 220 --------
9. 897 - 347 --------	10. 426 - 349 --------	11. 687 - 357 --------	12. 893 - 211 --------	13. 242 - 206 --------	14. 588 - 295 --------	15. 785 - 290 --------	16. 858 - 644 --------
17. 910 - 592 --------	18. 129 - 115 --------	19. 340 - 137 --------	20. 937 - 623 --------	21. 410 - 356 --------	22. 297 - 184 --------	23. 764 - 534 --------	24. 915 - 744 --------
25. 795 - 596 --------	26. 598 - 368 --------	27. 413 - 354 --------	28. 760 - 326 --------	29. 312 - 208 --------	30. 136 - 135 --------	31. 511 - 196 --------	32. 629 - 519 --------
33. 897 - 561 --------	34. 341 - 312 --------	35. 614 - 121 --------	36. 970 - 737 --------	37. 441 - 352 --------	38. 478 - 412 --------	39. 297 - 190 --------	40. 461 - 282 --------
41. 856 - 669 --------	42. 664 - 127 --------	43. 750 - 298 --------	44. 432 - 130 --------	45. 885 - 642 --------	46. 365 - 300 --------	47. 203 - 200 --------	48. 250 - 117 --------

Find the difference

Complete all the activities (Subtraction)

Name:................................ Date:................................

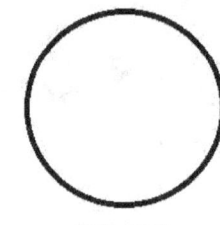

SCORE

1. 801 - 183 618	2. 685 - 216 469	3. 738 - 100 638	4. 669 - 299 370	5. 162 - 145 17	6. 366 - 145 221	7. 538 - 365 173	8. 613 - 220 393
9. 897 - 347 550	10. 426 - 349 77	11. 687 - 357 330	12. 893 - 211 682	13. 242 - 206 36	14. 588 - 295 293	15. 785 - 290 495	16. 858 - 644 214
17. 910 - 592 318	18. 129 - 115 14	19. 340 - 137 203	20. 937 - 623 314	21. 410 - 356 54	22. 297 - 184 113	23. 764 - 534 230	24. 915 - 744 171
25. 795 - 596 199	26. 598 - 368 230	27. 413 - 354 59	28. 760 - 326 434	29. 312 - 208 104	30. 136 - 135 1	31. 511 - 196 315	32. 629 - 519 110
33. 897 - 561 336	34. 341 - 312 29	35. 614 - 121 493	36. 970 - 737 233	37. 441 - 352 89	38. 478 - 412 66	39. 297 - 190 107	40. 461 - 282 179
41. 856 - 669 187	42. 664 - 127 537	43. 750 - 298 452	44. 432 - 130 302	45. 885 - 642 243	46. 365 - 300 65	47. 203 - 200 3	48. 250 - 117 133

Name:................................   Date:..

Find the difference

Complete all the activities (Subtraction)

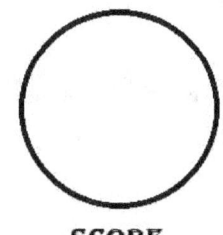

SCORE

1.	314	2.	587	3.	447	4.	296	5.	771	6.	243	7.	450	8.	211
	- 191		- 178		- 109		- 244		- 652		- 187		- 418		- 187

9.	199	10.	208	11.	645	12.	162	13.	289	14.	106	15.	770	16.	654
	- 153		- 176		- 395		- 136		- 133		- 105		- 242		- 327

17.	613	18.	629	19.	920	20.	207	21.	580	22.	371	23.	232	24.	731
	- 258		- 519		- 298		- 187		- 529		- 344		- 200		- 709

25.	610	26.	680	27.	863	28.	555	29.	422	30.	954	31.	117	32.	988
	- 362		- 436		- 453		- 417		- 409		- 279		- 106		- 574

33.	285	34.	135	35.	387	36.	881	37.	475	38.	253	39.	315	40.	490
	- 239		- 103		- 195		- 779		- 214		- 209		- 136		- 276

41.	527	42.	382	43.	610	44.	604	45.	821	46.	490	47.	362	48.	815
	- 118		- 165		- 119		- 155		- 742		- 471		- 338		- 241

Name:..................................   Date:..................................

Find the difference

Complete all the activities (Subtraction)

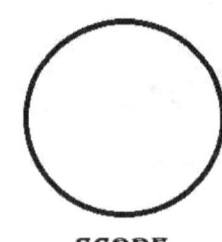

SCORE

1. 314 − 191 = 123	2. 587 − 178 = 409	3. 447 − 109 = 338	4. 296 − 244 = 52	5. 771 − 652 = 119	6. 243 − 187 = 56	7. 450 − 418 = 32	8. 211 − 187 = 24
9. 199 − 153 = 46	10. 208 − 176 = 32	11. 645 − 395 = 250	12. 162 − 136 = 26	13. 289 − 133 = 156	14. 106 − 105 = 1	15. 770 − 242 = 528	16. 654 − 327 = 327
17. 613 − 258 = 355	18. 629 − 519 = 110	19. 920 − 298 = 622	20. 207 − 187 = 20	21. 580 − 529 = 51	22. 371 − 344 = 27	23. 232 − 200 = 32	24. 731 − 709 = 22
25. 610 − 362 = 248	26. 680 − 436 = 244	27. 863 − 453 = 410	28. 555 − 417 = 138	29. 422 − 409 = 13	30. 954 − 279 = 675	31. 117 − 106 = 11	32. 988 − 574 = 414
33. 285 − 239 = 46	34. 135 − 103 = 32	35. 387 − 195 = 192	36. 881 − 779 = 102	37. 475 − 214 = 261	38. 253 − 209 = 44	39. 315 − 136 = 179	40. 490 − 276 = 214
41. 527 − 118 = 409	42. 382 − 165 = 217	43. 610 − 119 = 491	44. 604 − 155 = 449	45. 821 − 742 = 79	46. 490 − 471 = 19	47. 362 − 338 = 24	48. 815 − 241 = 574

Name:.................................   Date:.................................

Find the difference

Complete all the activities (Subtraction)

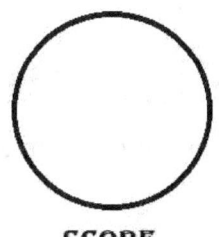

SCORE

1. 914 - 725 --------	2. 726 - 349 --------	3. 747 - 501 --------	4. 681 - 237 --------	5. 680 - 550 --------	6. 246 - 179 --------	7. 363 - 309 --------	8. 512 - 179 --------
9. 574 - 337 --------	10. 760 - 746 --------	11. 121 - 116 --------	12. 968 - 246 --------	13. 270 - 258 --------	14. 351 - 208 --------	15. 907 - 840 --------	16. 619 - 363 --------
17. 558 - 546 --------	18. 824 - 441 --------	19. 524 - 115 --------	20. 757 - 662 --------	21. 456 - 127 --------	22. 627 - 517 --------	23. 886 - 113 --------	24. 833 - 105 --------
25. 488 - 194 --------	26. 482 - 403 --------	27. 258 - 229 --------	28. 256 - 123 --------	29. 821 - 481 --------	30. 288 - 248 --------	31. 406 - 166 --------	32. 405 - 322 --------
33. 278 - 217 --------	34. 131 - 110 --------	35. 989 - 978 --------	36. 747 - 417 --------	37. 791 - 707 --------	38. 624 - 596 --------	39. 253 - 127 --------	40. 364 - 146 --------
41. 203 - 128 --------	42. 997 - 687 --------	43. 543 - 448 --------	44. 398 - 306 --------	45. 819 - 572 --------	46. 701 - 667 --------	47. 844 - 800 --------	48. 563 - 183 --------

Name:................................ Date:.................................

Find the difference

Complete all the activities (Subtraction)

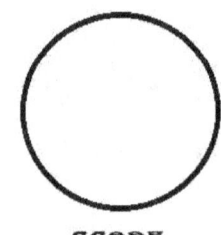

SCORE

| 1. | 914
- 725
189 | 2. | 726
- 349
377 | 3. | 747
- 501
246 | 4. | 681
- 237
444 | 5. | 680
- 550
130 | 6. | 246
- 179
67 | 7. | 363
- 309
54 | 8. | 512
- 179
333 |

| 9. | 574
- 337
237 | 10. | 760
- 746
14 | 11. | 121
- 116
5 | 12. | 968
- 246
722 | 13. | 270
- 258
12 | 14. | 351
- 208
143 | 15. | 907
- 840
67 | 16. | 619
- 363
256 |

| 17. | 558
- 546
12 | 18. | 824
- 441
383 | 19. | 524
- 115
409 | 20. | 757
- 662
95 | 21. | 456
- 127
329 | 22. | 627
- 517
110 | 23. | 886
- 113
773 | 24. | 833
- 105
728 |

| 25. | 488
- 194
294 | 26. | 482
- 403
79 | 27. | 258
- 229
29 | 28. | 256
- 123
133 | 29. | 821
- 481
340 | 30. | 288
- 248
40 | 31. | 406
- 166
240 | 32. | 405
- 322
83 |

| 33. | 278
- 217
61 | 34. | 131
- 110
21 | 35. | 989
- 978
11 | 36. | 747
- 417
330 | 37. | 791
- 707
84 | 38. | 624
- 596
28 | 39. | 253
- 127
126 | 40. | 364
- 146
218 |

| 41. | 203
- 128
75 | 42. | 997
- 687
310 | 43. | 543
- 448
95 | 44. | 398
- 306
92 | 45. | 819
- 572
247 | 46. | 701
- 667
34 | 47. | 844
- 800
44 | 48. | 563
- 183
380 |

Find the difference

Complete all the activities (Subtraction)

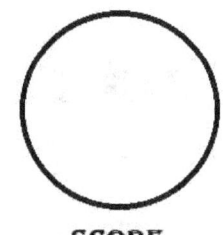

SCORE

1.   763
  - 669
	2. 633
	  - 217
		3. 983
		  - 538
			4. 116
			  - 107
				5. 570
				  - 539
					6. 121
					  - 106
						7. 311
						  - 148
							8. 980
							  - 653

9.   311
  - 108
	10. 707
	  - 141
		11. 297
		  - 235
			12. 312
			  - 188
				13. 391
				  - 200
					14. 567
					  - 255
						15. 387
						  - 323
							16. 696
							  - 593

17.   427
  - 232
	18. 288
	  - 159
		19. 622
		  - 139
			20. 272
			  - 206
				21. 329
				  - 227
					22. 218
					  - 166
						23. 785
						  - 124
							24. 511
							  - 335

25.   710
  - 125
	26. 737
	  - 604
		27. 339
		  - 207
			28. 963
			  - 145
				29. 660
				  - 201
					30. 906
					  - 615
						31. 978
						  - 702
							32. 213
							  - 184

33.   343
  - 147
	34. 938
	  - 797
		35. 116
		  - 111
			36. 143
			  - 107
				37. 860
				  - 143
					38. 831
					  - 586
						39. 598
						  - 437
							40. 686
							  - 269

41.   896
  - 854
	42. 688
	  - 263
		43. 161
		  - 153
			44. 605
			  - 563
				45. 499
				  - 162
					46. 720
					  - 348
						47. 260
						  - 188
							48. 677
							  - 428

Find the difference

Complete all the activities (Subtraction)

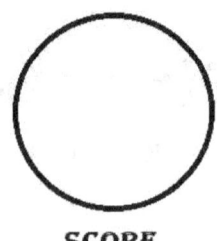

SCORE

1. 763	2. 633	3. 983	4. 116	5. 570	6. 121	7. 311	8. 980
- 669	- 217	- 538	- 107	- 539	- 106	- 148	- 653
94	416	445	9	31	15	163	327

9. 311	10. 707	11. 297	12. 312	13. 391	14. 567	15. 387	16. 696
- 108	- 141	- 235	- 188	- 200	- 255	- 323	- 593
203	566	62	124	191	312	64	103

17. 427	18. 288	19. 622	20. 272	21. 329	22. 218	23. 785	24. 511
- 232	- 159	- 139	- 206	- 227	- 166	- 124	- 335
195	129	483	66	102	52	661	176

25. 710	26. 737	27. 339	28. 963	29. 660	30. 906	31. 978	32. 213
- 125	- 604	- 207	- 145	- 201	- 615	- 702	- 184
585	133	132	818	459	291	276	29

33. 343	34. 938	35. 116	36. 143	37. 860	38. 831	39. 598	40. 686
- 147	- 797	- 111	- 107	- 143	- 586	- 437	- 269
196	141	5	36	717	245	161	417

41. 896	42. 688	43. 161	44. 605	45. 499	46. 720	47. 260	48. 677
- 854	- 263	- 153	- 563	- 162	- 348	- 188	- 428
42	425	8	42	337	372	72	249

Name:................................. Date:.................................

Find the difference

Complete all the activities (Subtraction)

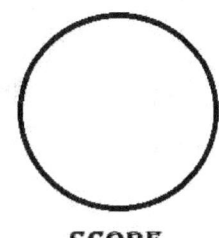

SCORE

1. 485 - 248 --------	2. 358 - 246 --------	3. 192 - 158 --------	4. 598 - 106 --------	5. 993 - 589 --------	6. 950 - 278 --------	7. 670 - 268 --------	8. 548 - 523 --------
9. 575 - 389 --------	10. 778 - 120 --------	11. 129 - 125 --------	12. 711 - 343 --------	13. 669 - 109 --------	14. 449 - 401 --------	15. 977 - 668 --------	16. 984 - 360 --------
17. 473 - 192 --------	18. 313 - 136 --------	19. 451 - 130 --------	20. 267 - 143 --------	21. 453 - 327 --------	22. 356 - 233 --------	23. 934 - 394 --------	24. 355 - 138 --------
25. 691 - 541 --------	26. 884 - 546 --------	27. 816 - 539 --------	28. 582 - 380 --------	29. 734 - 664 --------	30. 939 - 173 --------	31. 314 - 248 --------	32. 132 - 111 --------
33. 608 - 431 --------	34. 815 - 378 --------	35. 153 - 117 --------	36. 599 - 322 --------	37. 480 - 288 --------	38. 672 - 323 --------	39. 605 - 181 --------	40. 523 - 219 --------
41. 758 - 258 --------	42. 284 - 142 --------	43. 797 - 388 --------	44. 547 - 458 --------	45. 738 - 518 --------	46. 517 - 309 --------	47. 236 - 113 --------	48. 119 - 113 --------

Name:..................................   Date:..................................

Find the difference

Complete all the activities (Subtraction)

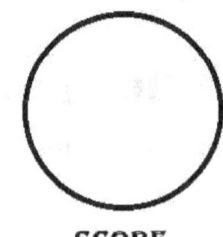

SCORE

1. 485 - 248 237	2. 358 - 246 112	3. 192 - 158 34	4. 598 - 106 492	5. 993 - 589 404	6. 950 - 278 672	7. 670 - 268 402	8. 548 - 523 25
9. 575 - 389 186	10. 778 - 120 658	11. 129 - 125 4	12. 711 - 343 368	13. 669 - 109 560	14. 449 - 401 48	15. 977 - 668 309	16. 984 - 360 624
17. 473 - 192 281	18. 313 - 136 177	19. 451 - 130 321	20. 267 - 143 124	21. 453 - 327 126	22. 356 - 233 123	23. 934 - 394 540	24. 355 - 138 217
25. 691 - 541 150	26. 884 - 546 338	27. 816 - 539 277	28. 582 - 380 202	29. 734 - 664 70	30. 939 - 173 766	31. 314 - 248 66	32. 132 - 111 21
33. 608 - 431 177	34. 815 - 378 437	35. 153 - 117 36	36. 599 - 322 277	37. 480 - 288 192	38. 672 - 323 349	39. 605 - 181 424	40. 523 - 219 304
41. 758 - 258 500	42. 284 - 142 142	43. 797 - 388 409	44. 547 - 458 89	45. 738 - 518 220	46. 517 - 309 208	47. 236 - 113 123	48. 119 - 113 6

Name:................................ Date:................................

Find the difference

Complete all the activities (Subtraction)

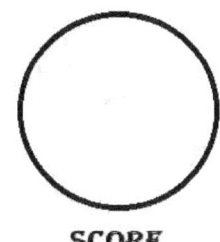

1. 557 - 123	2. 713 - 478	3. 144 - 107	4. 362 - 117	5. 326 - 112	6. 237 - 197	7. 997 - 625	8. 588 - 573
9. 167 - 121	10. 105 - 103	11. 479 - 374	12. 267 - 146	13. 597 - 244	14. 304 - 256	15. 683 - 294	16. 121 - 103
17. 280 - 234	18. 279 - 249	19. 859 - 327	20. 330 - 277	21. 524 - 104	22. 264 - 221	23. 357 - 343	24. 326 - 310
25. 389 - 174	26. 103 - 101	27. 371 - 213	28. 908 - 735	29. 877 - 418	30. 157 - 143	31. 948 - 548	32. 475 - 444
33. 127 - 127	34. 558 - 217	35. 448 - 347	36. 212 - 187	37. 156 - 113	38. 196 - 140	39. 585 - 317	40. 988 - 639
41. 135 - 110	42. 497 - 256	43. 397 - 273	44. 868 - 702	45. 859 - 413	46. 970 - 479	47. 312 - 205	48. 844 - 802

Name:................................   Date:................................

Find the difference

Complete all the activities (Subtraction)

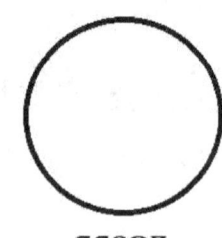

SCORE

1. 557 - 123 434	2. 713 - 478 235	3. 144 - 107 37	4. 362 - 117 245	5. 326 - 112 214	6. 237 - 197 40	7. 997 - 625 372	8. 588 - 573 15
9. 167 - 121 46	10. 105 - 103 2	11. 479 - 374 105	12. 267 - 146 121	13. 597 - 244 353	14. 304 - 256 48	15. 683 - 294 389	16. 121 - 103 18
17. 280 - 234 46	18. 279 - 249 30	19. 859 - 327 532	20. 330 - 277 53	21. 524 - 104 420	22. 264 - 221 43	23. 357 - 343 14	24. 326 - 310 16
25. 389 - 174 215	26. 103 - 101 2	27. 371 - 213 158	28. 908 - 735 173	29. 877 - 418 459	30. 157 - 143 14	31. 948 - 548 400	32. 475 - 444 31
33. 127 - 127 0	34. 558 - 217 341	35. 448 - 347 101	36. 212 - 187 25	37. 156 - 113 43	38. 196 - 140 56	39. 585 - 317 268	40. 988 - 639 349
41. 135 - 110 25	42. 497 - 256 241	43. 397 - 273 124	44. 868 - 702 166	45. 859 - 413 446	46. 970 - 479 491	47. 312 - 205 107	48. 844 - 802 42

Find the difference

Complete all the activities (Subtraction)

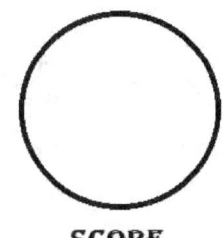

SCORE

| 1. | 632
- 241
-------- | 2. | 737
- 284
-------- | 3. | 493
- 491
-------- | 4. | 587
- 398
-------- | 5. | 217
- 127
-------- | 6. | 439
- 122
-------- | 7. | 379
- 236
-------- | 8. | 321
- 192
-------- |

| 9. | 478
- 273
-------- | 10. | 748
- 518
-------- | 11. | 178
- 102
-------- | 12. | 629
- 336
-------- | 13. | 228
- 193
-------- | 14. | 469
- 460
-------- | 15. | 768
- 560
-------- | 16. | 586
- 375
-------- |

| 17. | 467
- 160
-------- | 18. | 838
- 303
-------- | 19. | 764
- 564
-------- | 20. | 252
- 106
-------- | 21. | 779
- 664
-------- | 22. | 337
- 124
-------- | 23. | 286
- 247
-------- | 24. | 526
- 253
-------- |

| 25. | 250
- 182
-------- | 26. | 114
- 111
-------- | 27. | 797
- 227
-------- | 28. | 870
- 538
-------- | 29. | 709
- 259
-------- | 30. | 232
- 194
-------- | 31. | 626
- 399
-------- | 32. | 880
- 444
-------- |

| 33. | 718
- 601
-------- | 34. | 692
- 187
-------- | 35. | 698
- 681
-------- | 36. | 711
- 541
-------- | 37. | 359
- 276
-------- | 38. | 900
- 305
-------- | 39. | 782
- 431
-------- | 40. | 316
- 176
-------- |

| 41. | 327
- 112
-------- | 42. | 826
- 241
-------- | 43. | 706
- 562
-------- | 44. | 811
- 261
-------- | 45. | 376
- 170
-------- | 46. | 637
- 356
-------- | 47. | 597
- 327
-------- | 48. | 511
- 177
-------- |

Name:................................ Date:....................................

Find the difference

Complete all the activities (Subtraction)

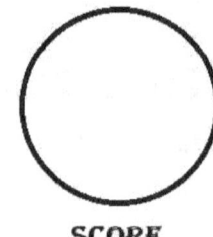

SCORE

1. 632	2. 737	3. 493	4. 587	5. 217	6. 439	7. 379	8. 321
- 241	- 284	- 491	- 398	- 127	- 122	- 236	- 192
391	453	2	189	90	317	143	129

9. 478	10. 748	11. 178	12. 629	13. 228	14. 469	15. 768	16. 586
- 273	- 518	- 102	- 336	- 193	- 460	- 560	- 375
205	230	76	293	35	9	208	211

17. 467	18. 838	19. 764	20. 252	21. 779	22. 337	23. 286	24. 526
- 160	- 303	- 564	- 106	- 664	- 124	- 247	- 253
307	535	200	146	115	213	39	273

25. 250	26. 114	27. 797	28. 870	29. 709	30. 232	31. 626	32. 880
- 182	- 111	- 227	- 538	- 259	- 194	- 399	- 444
68	3	570	332	450	38	227	436

33. 718	34. 692	35. 698	36. 711	37. 359	38. 900	39. 782	40. 316
- 601	- 187	- 681	- 541	- 276	- 305	- 431	- 176
117	505	17	170	83	595	351	140

41. 327	42. 826	43. 706	44. 811	45. 376	46. 637	47. 597	48. 511
- 112	- 241	- 562	- 261	- 170	- 356	- 327	- 177
215	585	144	550	206	281	270	334

Name:................................   Date:................................

Find the difference

Complete all the activities (Subtraction)

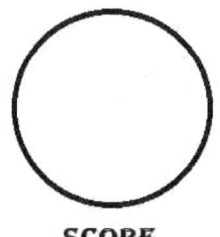

SCORE

1.	966 - 813	2.	473 - 248	3.	217 - 138	4.	723 - 221	5.	882 - 330	6.	540 - 142	7.	362 - 257	8.	828 - 239

9.	119 - 115	10.	514 - 320	11.	783 - 560	12.	789 - 737	13.	118 - 108	14.	854 - 411	15.	436 - 168	16.	251 - 226

17.	299 - 240	18.	364 - 347	19.	164 - 154	20.	302 - 287	21.	424 - 299	22.	392 - 211	23.	345 - 270	24.	940 - 694

25.	498 - 262	26.	509 - 411	27.	996 - 929	28.	485 - 211	29.	151 - 125	30.	438 - 336	31.	539 - 223	32.	837 - 592

33.	213 - 174	34.	887 - 564	35.	376 - 341	36.	208 - 119	37.	611 - 373	38.	255 - 123	39.	414 - 221	40.	686 - 291

41.	547 - 159	42.	972 - 944	43.	772 - 374	44.	851 - 323	45.	346 - 101	46.	890 - 559	47.	522 - 322	48.	493 - 131

Find the difference

Complete all the activities (Subtraction)

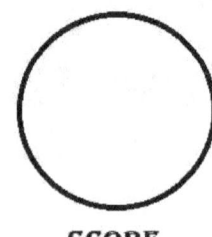

SCORE

1. 966 - 813 153	2. 473 - 248 225	3. 217 - 138 79	4. 723 - 221 502	5. 882 - 330 552	6. 540 - 142 398	7. 362 - 257 105	8. 828 - 239 589
9. 119 - 115 4	10. 514 - 320 194	11. 783 - 560 223	12. 789 - 737 52	13. 118 - 108 10	14. 854 - 411 443	15. 436 - 168 268	16. 251 - 226 25
17. 299 - 240 59	18. 364 - 347 17	19. 164 - 154 10	20. 302 - 287 15	21. 424 - 299 125	22. 392 - 211 181	23. 345 - 270 75	24. 940 - 694 246
25. 498 - 262 236	26. 509 - 411 98	27. 996 - 929 67	28. 485 - 211 274	29. 151 - 125 26	30. 438 - 336 102	31. 539 - 223 316	32. 837 - 592 245
33. 213 - 174 39	34. 887 - 564 323	35. 376 - 341 35	36. 208 - 119 89	37. 611 - 373 238	38. 255 - 123 132	39. 414 - 221 193	40. 686 - 291 395
41. 547 - 159 388	42. 972 - 944 28	43. 772 - 374 398	44. 851 - 323 528	45. 346 - 101 245	46. 890 - 559 331	47. 522 - 322 200	48. 493 - 131 362

Name:.................................. Date:..................................

Find the difference

Complete all the activities (Subtraction)

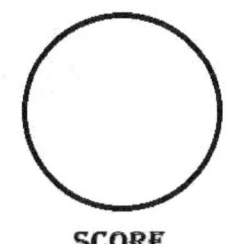

SCORE

| 1. 813
- 569
-------- | 2. 328
- 240
-------- | 3. 837
- 354
-------- | 4. 460
- 305
-------- | 5. 357
- 172
-------- | 6. 273
- 180
-------- | 7. 777
- 291
-------- | 8. 945
- 666
-------- |

| 9. 853
- 138
-------- | 10. 279
- 164
-------- | 11. 379
- 338
-------- | 12. 927
- 694
-------- | 13. 840
- 528
-------- | 14. 385
- 186
-------- | 15. 995
- 583
-------- | 16. 736
- 126
-------- |

| 17. 188
- 108
-------- | 18. 418
- 320
-------- | 19. 219
- 104
-------- | 20. 555
- 170
-------- | 21. 758
- 364
-------- | 22. 713
- 387
-------- | 23. 294
- 172
-------- | 24. 589
- 585
-------- |

| 25. 157
- 112
-------- | 26. 768
- 380
-------- | 27. 394
- 124
-------- | 28. 791
- 750
-------- | 29. 763
- 455
-------- | 30. 812
- 271
-------- | 31. 125
- 102
-------- | 32. 615
- 484
-------- |

| 33. 614
- 178
-------- | 34. 427
- 316
-------- | 35. 935
- 307
-------- | 36. 583
- 487
-------- | 37. 589
- 123
-------- | 38. 660
- 609
-------- | 39. 253
- 169
-------- | 40. 552
- 303
-------- |

| 41. 131
- 110
-------- | 42. 315
- 269
-------- | 43. 816
- 306
-------- | 44. 741
- 110
-------- | 45. 391
- 220
-------- | 46. 958
- 612
-------- | 47. 320
- 158
-------- | 48. 213
- 172
-------- |

Name:................................... Date:......................................

Find the difference

Complete all the activities (Subtraction)

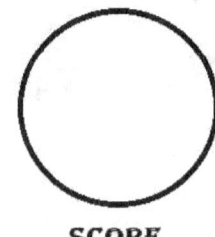

SCORE

| 1. | 813
- 569
244 | 2. | 328
- 240
88 | 3. | 837
- 354
483 | 4. | 460
- 305
155 | 5. | 357
- 172
185 | 6. | 273
- 180
93 | 7. | 777
- 291
486 | 8. | 945
- 666
279 |

| 9. | 853
- 138
715 | 10. | 279
- 164
115 | 11. | 379
- 338
41 | 12. | 927
- 694
233 | 13. | 840
- 528
312 | 14. | 385
- 186
199 | 15. | 995
- 583
412 | 16. | 736
- 126
610 |

| 17. | 188
- 108
80 | 18. | 418
- 320
98 | 19. | 219
- 104
115 | 20. | 555
- 170
385 | 21. | 758
- 364
394 | 22. | 713
- 387
326 | 23. | 294
- 172
122 | 24. | 589
- 585
4 |

| 25. | 157
- 112
45 | 26. | 768
- 380
388 | 27. | 394
- 124
270 | 28. | 791
- 750
41 | 29. | 763
- 455
308 | 30. | 812
- 271
541 | 31. | 125
- 102
23 | 32. | 615
- 484
131 |

| 33. | 614
- 178
436 | 34. | 427
- 316
111 | 35. | 935
- 307
628 | 36. | 583
- 487
96 | 37. | 589
- 123
466 | 38. | 660
- 609
51 | 39. | 253
- 169
84 | 40. | 552
- 303
249 |

| 41. | 131
- 110
21 | 42. | 315
- 269
46 | 43. | 816
- 306
510 | 44. | 741
- 110
631 | 45. | 391
- 220
171 | 46. | 958
- 612
346 | 47. | 320
- 158
162 | 48. | 213
- 172
41 |

Name:.................................. Date:..............................

Find the difference

Complete all the activities (Subtraction)

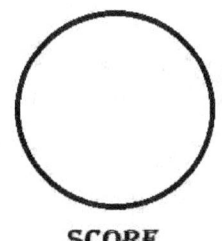

SCORE

| 1. 844
- 796
-------- | 2. 907
- 151
-------- | 3. 109
- 108
-------- | 4. 156
- 146
-------- | 5. 704
- 669
-------- | 6. 887
- 321
-------- | 7. 512
- 454
-------- | 8. 793
- 299
-------- |

| 9. 130
- 110
-------- | 10. 285
- 230
-------- | 11. 871
- 706
-------- | 12. 954
- 342
-------- | 13. 494
- 237
-------- | 14. 402
- 256
-------- | 15. 364
- 348
-------- | 16. 600
- 239
-------- |

| 17. 928
- 918
-------- | 18. 174
- 117
-------- | 19. 598
- 319
-------- | 20. 495
- 160
-------- | 21. 336
- 200
-------- | 22. 204
- 189
-------- | 23. 471
- 260
-------- | 24. 602
- 333
-------- |

| 25. 676
- 484
-------- | 26. 607
- 557
-------- | 27. 154
- 123
-------- | 28. 800
- 129
-------- | 29. 986
- 644
-------- | 30. 662
- 472
-------- | 31. 593
- 369
-------- | 32. 568
- 442
-------- |

| 33. 446
- 307
-------- | 34. 424
- 252
-------- | 35. 798
- 475
-------- | 36. 759
- 491
-------- | 37. 912
- 144
-------- | 38. 495
- 271
-------- | 39. 704
- 158
-------- | 40. 883
- 105
-------- |

| 41. 810
- 718
-------- | 42. 887
- 593
-------- | 43. 439
- 322
-------- | 44. 296
- 198
-------- | 45. 289
- 108
-------- | 46. 323
- 310
-------- | 47. 832
- 670
-------- | 48. 641
- 488
-------- |

Name:................................... Date:...................................

Find the difference

Complete all the activities (Subtraction)

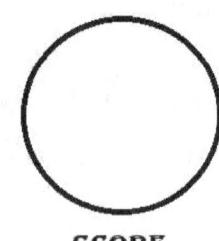

SCORE

1. 844 - 796 48	2. 907 - 151 756	3. 109 - 108 1	4. 156 - 146 10	5. 704 - 669 35	6. 887 - 321 566	7. 512 - 454 58	8. 793 - 299 494
9. 130 - 110 20	10. 285 - 230 55	11. 871 - 706 165	12. 954 - 342 612	13. 494 - 237 257	14. 402 - 256 146	15. 364 - 348 16	16. 600 - 239 361
17. 928 - 918 10	18. 174 - 117 57	19. 598 - 319 279	20. 495 - 160 335	21. 336 - 200 136	22. 204 - 189 15	23. 471 - 260 211	24. 602 - 333 269
25. 676 - 484 192	26. 607 - 557 50	27. 154 - 123 31	28. 800 - 129 671	29. 986 - 644 342	30. 662 - 472 190	31. 593 - 369 224	32. 568 - 442 126
33. 446 - 307 139	34. 424 - 252 172	35. 798 - 475 323	36. 759 - 491 268	37. 912 - 144 768	38. 495 - 271 224	39. 704 - 158 546	40. 883 - 105 778
41. 810 - 718 92	42. 887 - 593 294	43. 439 - 322 117	44. 296 - 198 98	45. 289 - 108 181	46. 323 - 310 13	47. 832 - 670 162	48. 641 - 488 153

Find the difference

Complete all the activities (Subtraction)

Name:...................................   Date:...................................

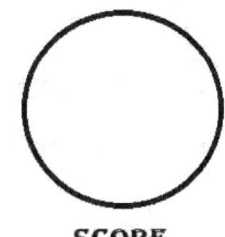

SCORE

1. 912 - 412	2. 521 - 443	3. 165 - 106	4. 719 - 165	5. 940 - 915	6. 997 - 105	7. 984 - 406	8. 432 - 395
9. 768 - 513	10. 884 - 632	11. 703 - 639	12. 419 - 333	13. 441 - 394	14. 334 - 272	15. 359 - 305	16. 800 - 188
17. 507 - 394	18. 576 - 237	19. 664 - 131	20. 941 - 690	21. 264 - 226	22. 722 - 353	23. 691 - 226	24. 563 - 500
25. 704 - 317	26. 277 - 135	27. 914 - 852	28. 142 - 100	29. 127 - 120	30. 492 - 182	31. 844 - 809	32. 834 - 203
33. 271 - 249	34. 597 - 230	35. 834 - 160	36. 336 - 262	37. 714 - 205	38. 545 - 500	39. 282 - 272	40. 166 - 111
41. 412 - 224	42. 157 - 153	43. 636 - 412	44. 297 - 204	45. 707 - 416	46. 331 - 305	47. 657 - 504	48. 886 - 281

Name:.................................... Date:....................................

Find the difference

Complete all the activities (Subtraction)

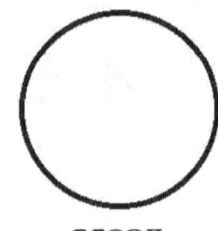

SCORE

| 1. | 912
- 412
500 | 2. | 521
- 443
78 | 3. | 165
- 106
59 | 4. | 719
- 165
554 | 5. | 940
- 915
25 | 6. | 997
- 105
892 | 7. | 984
- 406
578 | 8. | 432
- 395
37 |

| 9. | 768
- 513
255 | 10. | 884
- 632
252 | 11. | 703
- 639
64 | 12. | 419
- 333
86 | 13. | 441
- 394
47 | 14. | 334
- 272
62 | 15. | 359
- 305
54 | 16. | 800
- 188
612 |

| 17. | 507
- 394
113 | 18. | 576
- 237
339 | 19. | 664
- 131
533 | 20. | 941
- 690
251 | 21. | 264
- 226
38 | 22. | 722
- 353
369 | 23. | 691
- 226
465 | 24. | 563
- 500
63 |

| 25. | 704
- 317
387 | 26. | 277
- 135
142 | 27. | 914
- 852
62 | 28. | 142
- 100
42 | 29. | 127
- 120
7 | 30. | 492
- 182
310 | 31. | 844
- 809
35 | 32. | 834
- 203
631 |

| 33. | 271
- 249
22 | 34. | 597
- 230
367 | 35. | 834
- 160
674 | 36. | 336
- 262
74 | 37. | 714
- 205
509 | 38. | 545
- 500
45 | 39. | 282
- 272
10 | 40. | 166
- 111
55 |

| 41. | 412
- 224
188 | 42. | 157
- 153
4 | 43. | 636
- 412
224 | 44. | 297
- 204
93 | 45. | 707
- 416
291 | 46. | 331
- 305
26 | 47. | 657
- 504
153 | 48. | 886
- 281
605 |

Name:................................ Date:................................

Find the difference

Complete all the activities (Subtraction)

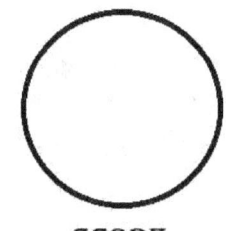

SCORE

1. 959 - 574 -------	2. 295 - 253 -------	3. 693 - 250 -------	4. 143 - 125 -------	5. 186 - 144 -------

1. 959
   - 574
   ----
2. 295
   - 253
   ----
3. 693
   - 250
   ----
4. 143
   - 125
   ----
5. 186
   - 144
   ----
6. 562
   - 440
   ----
7. 485
   - 157
   ----
8. 826
   - 160
   ----

9. 698
   - 261
   ----
10. 875
   - 434
   ----
11. 899
   - 808
   ----
12. 589
   - 398
   ----
13. 688
   - 114
   ----
14. 627
   - 256
   ----
15. 842
   - 378
   ----
16. 161
   - 158
   ----

17. 367
   - 171
   ----
18. 949
   - 652
   ----
19. 757
   - 290
   ----
20. 556
   - 143
   ----
21. 413
   - 249
   ----
22. 849
   - 165
   ----
23. 161
   - 130
   ----
24. 190
   - 147
   ----

25. 741
   - 212
   ----
26. 300
   - 170
   ----
27. 876
   - 443
   ----
28. 407
   - 121
   ----
29. 106
   - 104
   ----
30. 217
   - 174
   ----
31. 699
   - 550
   ----
32. 787
   - 596
   ----

33. 497
   - 321
   ----
34. 494
   - 164
   ----
35. 283
   - 129
   ----
36. 416
   - 219
   ----
37. 785
   - 626
   ----
38. 175
   - 131
   ----
39. 147
   - 121
   ----
40. 586
   - 182
   ----

41. 931
   - 149
   ----
42. 624
   - 441
   ----
43. 236
   - 198
   ----
44. 185
   - 173
   ----
45. 931
   - 827
   ----
46. 679
   - 171
   ----
47. 243
   - 225
   ----
48. 485
   - 330

Name:................................ Date:...............................

Find the difference

Complete all the activities (Subtraction)

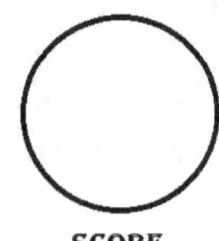

SCORE

| 1. | 959
- 574
385 | 2. | 295
- 253
42 | 3. | 693
- 250
443 | 4. | 143
- 125
18 | 5. | 186
- 144
42 | 6. | 562
- 440
122 | 7. | 485
- 157
328 | 8. | 826
- 160
666 |

| 9. | 698
- 261
437 | 10. | 875
- 434
441 | 11. | 899
- 808
91 | 12. | 589
- 398
191 | 13. | 688
- 114
574 | 14. | 627
- 256
371 | 15. | 842
- 378
464 | 16. | 161
- 158
3 |

| 17. | 367
- 171
196 | 18. | 949
- 652
297 | 19. | 757
- 290
467 | 20. | 556
- 143
413 | 21. | 413
- 249
164 | 22. | 849
- 165
684 | 23. | 161
- 130
31 | 24. | 190
- 147
43 |

| 25. | 741
- 212
529 | 26. | 300
- 170
130 | 27. | 876
- 443
433 | 28. | 407
- 121
286 | 29. | 106
- 104
2 | 30. | 217
- 174
43 | 31. | 699
- 550
149 | 32. | 787
- 596
191 |

| 33. | 497
- 321
176 | 34. | 494
- 164
330 | 35. | 283
- 129
154 | 36. | 416
- 219
197 | 37. | 785
- 626
159 | 38. | 175
- 131
44 | 39. | 147
- 121
26 | 40. | 586
- 182
404 |

| 41. | 931
- 149
782 | 42. | 624
- 441
183 | 43. | 236
- 198
38 | 44. | 185
- 173
12 | 45. | 931
- 827
104 | 46. | 679
- 171
508 | 47. | 243
- 225
18 | 48. | 485
- 330
155 |

Find the difference

Complete all the activities (Subtraction)

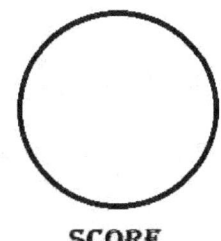

SCORE

1.	738 - 171	2.	680 - 397	3.	141 - 138	4.	390 - 261	5.	840 - 727	6.	629 - 150	7.	423 - 400	8.	115 - 108

9.	161 - 159	10.	739 - 488	11.	251 - 117	12.	613 - 472	13.	515 - 343	14.	315 - 171	15.	561 - 205	16.	295 - 274

17.	406 - 249	18.	733 - 696	19.	830 - 121	20.	211 - 210	21.	762 - 118	22.	843 - 326	23.	215 - 215	24.	545 - 543

25.	872 - 173	26.	150 - 110	27.	244 - 218	28.	485 - 329	29.	586 - 547	30.	145 - 121	31.	727 - 723	32.	139 - 123

33.	666 - 495	34.	779 - 693	35.	301 - 115	36.	763 - 126	37.	852 - 644	38.	883 - 211	39.	359 - 169	40.	636 - 348

41.	181 - 120	42.	523 - 277	43.	256 - 103	44.	756 - 441	45.	153 - 106	46.	459 - 420	47.	980 - 683	48.	361 - 294

Name:................................   Date:................................

Find the difference

Complete all the activities (Subtraction)

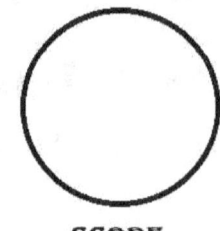

SCORE

1. 738	2. 680	3. 141	4. 390	5. 840	6. 629	7. 423	8. 115
- 171	- 397	- 138	- 261	- 727	- 150	- 400	- 108
567	283	3	129	113	479	23	7

9. 161	10. 739	11. 251	12. 613	13. 515	14. 315	15. 561	16. 295
- 159	- 488	- 117	- 472	- 343	- 171	- 205	- 274
2	251	134	141	172	144	356	21

17. 406	18. 733	19. 830	20. 211	21. 762	22. 843	23. 215	24. 545
- 249	- 696	- 121	- 210	- 118	- 326	- 215	- 543
157	37	709	1	644	517	0	2

25. 872	26. 150	27. 244	28. 485	29. 586	30. 145	31. 727	32. 139
- 173	- 110	- 218	- 329	- 547	- 121	- 723	- 123
699	40	26	156	39	24	4	16

33. 666	34. 779	35. 301	36. 763	37. 852	38. 883	39. 359	40. 636
- 495	- 693	- 115	- 126	- 644	- 211	- 169	- 348
171	86	186	637	208	672	190	288

41. 181	42. 523	43. 256	44. 756	45. 153	46. 459	47. 980	48. 361
- 120	- 277	- 103	- 441	- 106	- 420	- 683	- 294
61	246	153	315	47	39	297	67

Name:................................   Date:..................................

Find the difference

Complete all the activities (Subtraction)

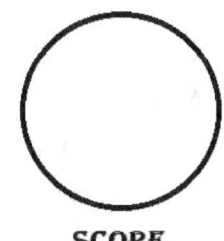

SCORE

1. 983 - 305 --------	2. 707 - 361 --------	3. 319 - 236 --------	4. 794 - 643 --------	5. 420 - 105 --------	6. 426 - 243 --------	7. 954 - 344 --------	8. 377 - 176 --------
9. 391 - 271 --------	10. 434 - 252 --------	11. 570 - 394 --------	12. 479 - 332 --------	13. 971 - 305 --------	14. 832 - 185 --------	15. 708 - 615 --------	16. 154 - 109 --------
17. 489 - 150 --------	18. 243 - 108 --------	19. 632 - 189 --------	20. 297 - 192 --------	21. 523 - 391 --------	22. 156 - 109 --------	23. 689 - 272 --------	24. 612 - 495 --------
25. 820 - 190 --------	26. 610 - 300 --------	27. 362 - 319 --------	28. 297 - 171 --------	29. 828 - 145 --------	30. 514 - 103 --------	31. 826 - 576 --------	32. 196 - 149 --------
33. 959 - 263 --------	34. 104 - 103 --------	35. 214 - 129 --------	36. 714 - 196 --------	37. 430 - 398 --------	38. 984 - 838 --------	39. 606 - 233 --------	40. 961 - 683 --------
41. 699 - 125 --------	42. 384 - 256 --------	43. 351 - 268 --------	44. 893 - 573 --------	45. 917 - 755 --------	46. 321 - 217 --------	47. 735 - 235 --------	48. 694 - 584 --------

Name:.................................   Date:.................................

Find the difference

Complete all the activities (Subtraction)

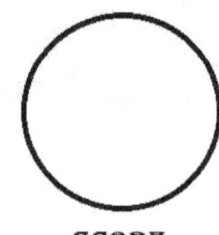

SCORE

1.	983	2.	707	3.	319	4.	794	5.	420	6.	426	7.	954	8.	377
	- 305		- 361		- 236		- 643		- 105		- 243		- 344		- 176
	678		346		83		151		315		183		610		201

9.	391	10.	434	11.	570	12.	479	13.	971	14.	832	15.	708	16.	154
	- 271		- 252		- 394		- 332		- 305		- 185		- 615		- 109
	120		182		176		147		666		647		93		45

17.	489	18.	243	19.	632	20.	297	21.	523	22.	156	23.	689	24.	612
	- 150		- 108		- 189		- 192		- 391		- 109		- 272		- 495
	339		135		443		105		132		47		417		117

25.	820	26.	610	27.	362	28.	297	29.	828	30.	514	31.	826	32.	196
	- 190		- 300		- 319		- 171		- 145		- 103		- 576		- 149
	630		310		43		126		683		411		250		47

33.	959	34.	104	35.	214	36.	714	37.	430	38.	984	39.	606	40.	961
	- 263		- 103		- 129		- 196		- 398		- 838		- 233		- 683
	696		1		85		518		32		146		373		278

41.	699	42.	384	43.	351	44.	893	45.	917	46.	321	47.	735	48.	694
	- 125		- 256		- 268		- 573		- 755		- 217		- 235		- 584
	574		128		83		320		162		104		500		110

Name:.. Date:...............................

Find the difference

Complete all the activities (Subtraction)

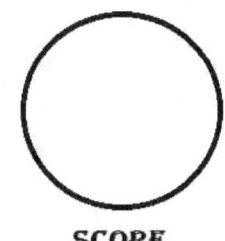

SCORE

1. 622 - 388 --------	2. 282 - 185 --------	3. 359 - 294 --------	4. 681 - 419 --------	5. 497 - 263 --------	6. 123 - 109 --------	7. 867 - 395 --------	8. 815 - 491 --------
9. 194 - 174 --------	10. 407 - 335 --------	11. 406 - 270 --------	12. 308 - 160 --------	13. 377 - 213 --------	14. 921 - 367 --------	15. 105 - 101 --------	16. 505 - 163 --------
17. 318 - 137 --------	18. 329 - 102 --------	19. 155 - 127 --------	20. 450 - 421 --------	21. 456 - 373 --------	22. 199 - 186 --------	23. 677 - 385 --------	24. 769 - 451 --------
25. 105 - 103 --------	26. 482 - 453 --------	27. 883 - 800 --------	28. 526 - 111 --------	29. 389 - 364 --------	30. 241 - 224 --------	31. 851 - 341 --------	32. 588 - 240 --------
33. 964 - 852 --------	34. 231 - 190 --------	35. 217 - 192 --------	36. 536 - 367 --------	37. 455 - 250 --------	38. 433 - 291 --------	39. 501 - 245 --------	40. 680 - 573 --------
41. 697 - 332 --------	42. 387 - 188 --------	43. 756 - 446 --------	44. 356 - 189 --------	45. 725 - 243 --------	46. 719 - 191 --------	47. 402 - 254 --------	48. 174 - 106 --------

Name:................................   Date:................................

Find the difference

Complete all the activities (Subtraction)

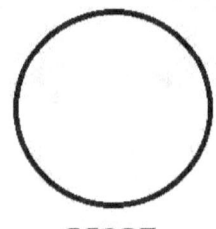

SCORE

| 1. | 622 − 388 = 234 | 2. | 282 − 185 = 97 | 3. | 359 − 294 = 65 | 4. | 681 − 419 = 262 | 5. | 497 − 263 = 234 | 6. | 123 − 109 = 14 | 7. | 867 − 395 = 472 | 8. | 815 − 491 = 324 |

1. 622 − 388 = 234
2. 282 − 185 = 97
3. 359 − 294 = 65
4. 681 − 419 = 262
5. 497 − 263 = 234
6. 123 − 109 = 14
7. 867 − 395 = 472
8. 815 − 491 = 324

9. 194 − 174 = 20
10. 407 − 335 = 72
11. 406 − 270 = 136
12. 308 − 160 = 148
13. 377 − 213 = 164
14. 921 − 367 = 554
15. 105 − 101 = 4
16. 505 − 163 = 342

17. 318 − 137 = 181
18. 329 − 102 = 227
19. 155 − 127 = 28
20. 450 − 421 = 29
21. 456 − 373 = 83
22. 199 − 186 = 13
23. 677 − 385 = 292
24. 769 − 451 = 318

25. 105 − 103 = 2
26. 482 − 453 = 29
27. 883 − 800 = 83
28. 526 − 111 = 415
29. 389 − 364 = 25
30. 241 − 224 = 17
31. 851 − 341 = 510
32. 588 − 240 = 348

33. 964 − 852 = 112
34. 231 − 190 = 41
35. 217 − 192 = 25
36. 536 − 367 = 169
37. 455 − 250 = 205
38. 433 − 291 = 142
39. 501 − 245 = 256
40. 680 − 573 = 107

41. 697 − 332 = 365
42. 387 − 188 = 199
43. 756 − 446 = 310
44. 356 − 189 = 167
45. 725 − 243 = 482
46. 719 − 191 = 528
47. 402 − 254 = 148
48. 174 − 106 = 68

Find the difference

Complete all the activities (Subtraction)

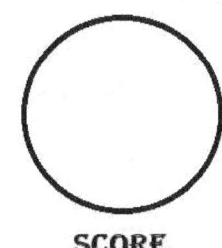

SCORE

1.   631
   - 627
	2. 342
	   - 267
		3. 880
		   - 387
			4. 262
			   - 184
				5. 790
				   - 541
					6. 953
					   - 912
						7. 897
						   - 219
							8. 949
							   - 389

9.   204
   - 107
	10. 784
	   - 159
		11. 866
		   - 159
			12. 496
			   - 417
				13. 768
				   - 738
					14. 487
					   - 166
						15. 145
						   - 102
							16. 655
							   - 351

17.   553
   - 253
	18. 117
	   - 111
		19. 609
		   - 589
			20. 742
			   - 730
				21. 577
				   - 201
					22. 871
					   - 124
						23. 677
						   - 383
							24. 907
							   - 287

25.   382
   - 343
	26. 571
	   - 543
		27. 144
		   - 115
			28. 662
			   - 389
				29. 597
				   - 333
					30. 953
					   - 878
						31. 943
						   - 108
							32. 519
							   - 399

33.   120
   - 110
	34. 503
	   - 321
		35. 217
		   - 135
			36. 796
			   - 208
				37. 418
				   - 348
					38. 599
					   - 420
						39. 297
						   - 181
							40. 244
							   - 136

41.   152
   - 110
	42. 871
	   - 246
		43. 577
		   - 423
			44. 462
			   - 162
				45. 741
				   - 424
					46. 187
					   - 109
						47. 111
						   - 109
							48. 139
							   - 120

Name:................................ Date:................................

Find the difference

Complete all the activities (Subtraction)

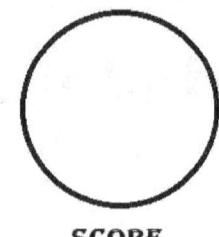

SCORE

| 1. | 631
- 627
4 | 2. | 342
- 267
75 | 3. | 880
- 387
493 | 4. | 262
- 184
78 | 5. | 790
- 541
249 | 6. | 953
- 912
41 | 7. | 897
- 219
678 | 8. | 949
- 389
560 |

9. 204   10. 784   11. 866   12. 496   13. 768   14. 487   15. 145   16. 655
   - 107   - 159   - 159   - 417   - 738   - 166   - 102   - 351
     97   625   707   79   30   321   43   304

17. 553   18. 117   19. 609   20. 742   21. 577   22. 871   23. 677   24. 907
   - 253   - 111   - 589   - 730   - 201   - 124   - 383   - 287
    300   6   20   12   376   747   294   620

25. 382   26. 571   27. 144   28. 662   29. 597   30. 953   31. 943   32. 519
   - 343   - 543   - 115   - 389   - 333   - 878   - 108   - 399
     39   28   29   273   264   75   835   120

33. 120   34. 503   35. 217   36. 796   37. 418   38. 599   39. 297   40. 244
   - 110   - 321   - 135   - 208   - 348   - 420   - 181   - 136
     10   182   82   588   70   179   116   108

41. 152   42. 871   43. 577   44. 462   45. 741   46. 187   47. 111   48. 139
   - 110   - 246   - 423   - 162   - 424   - 109   - 109   - 120
     42   625   154   300   317   78   2   19

Name:.................................. Date:...............................

Find the difference

Complete all the activities (Subtraction)

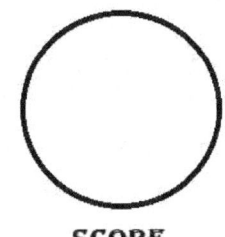

SCORE

| 1. 769
- 196
-------- | 2. 108
- 101
-------- | 3. 139
- 114
-------- | 4. 191
- 102
-------- | 5. 178
- 129
-------- | 6. 627
- 242
-------- | 7. 799
- 788
-------- | 8. 365
- 138
-------- |

| 9. 377
- 293
-------- | 10. 868
- 667
-------- | 11. 237
- 138
-------- | 12. 264
- 157
-------- | 13. 184
- 123
-------- | 14. 851
- 570
-------- | 15. 279
- 221
-------- | 16. 229
- 171
-------- |

| 17. 763
- 122
-------- | 18. 906
- 161
-------- | 19. 461
- 105
-------- | 20. 287
- 232
-------- | 21. 188
- 174
-------- | 22. 192
- 115
-------- | 23. 182
- 111
-------- | 24. 453
- 357
-------- |

| 25. 289
- 280
-------- | 26. 810
- 324
-------- | 27. 953
- 360
-------- | 28. 433
- 256
-------- | 29. 965
- 674
-------- | 30. 146
- 135
-------- | 31. 562
- 320
-------- | 32. 836
- 403
-------- |

| 33. 303
- 107
-------- | 34. 579
- 457
-------- | 35. 179
- 117
-------- | 36. 767
- 296
-------- | 37. 539
- 164
-------- | 38. 676
- 280
-------- | 39. 258
- 103
-------- | 40. 328
- 319
-------- |

| 41. 895
- 185
-------- | 42. 474
- 192
-------- | 43. 897
- 760
-------- | 44. 641
- 334
-------- | 45. 696
- 356
-------- | 46. 202
- 175
-------- | 47. 966
- 483
-------- | 48. 275
- 209
-------- |

Name:................................ Date:................................

Find the difference

Complete all the activities (Subtraction)

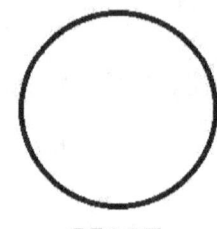

SCORE

1. 769 - 196 573	2. 108 - 101 7	3. 139 - 114 25	4. 191 - 102 89	5. 178 - 129 49	6. 627 - 242 385	7. 799 - 788 11	8. 365 - 138 227
9. 377 - 293 84	10. 868 - 667 201	11. 237 - 138 99	12. 264 - 157 107	13. 184 - 123 61	14. 851 - 570 281	15. 279 - 221 58	16. 229 - 171 58
17. 763 - 122 641	18. 906 - 161 745	19. 461 - 105 356	20. 287 - 232 55	21. 188 - 174 14	22. 192 - 115 77	23. 182 - 111 71	24. 453 - 357 96
25. 289 - 280 9	26. 810 - 324 486	27. 953 - 360 593	28. 433 - 256 177	29. 965 - 674 291	30. 146 - 135 11	31. 562 - 320 242	32. 836 - 403 433
33. 303 - 107 196	34. 579 - 457 122	35. 179 - 117 62	36. 767 - 296 471	37. 539 - 164 375	38. 676 - 280 396	39. 258 - 103 155	40. 328 - 319 9
41. 895 - 185 710	42. 474 - 192 282	43. 897 - 760 137	44. 641 - 334 307	45. 696 - 356 340	46. 202 - 175 27	47. 966 - 483 483	48. 275 - 209 66

Name:................................ Date:.................................

Find the difference
Complete all the activities (Subtraction)

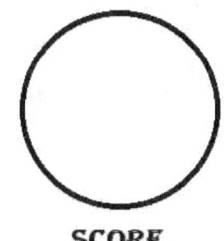

1. 118 - 112	2. 942 - 147	3. 640 - 450	4. 676 - 670	5. 291 - 195	6. 843 - 697	7. 697 - 134	8. 362 - 347
--------	--------	--------	--------	--------	--------	--------	--------

9. 469 - 134	10. 893 - 423	11. 153 - 130	12. 572 - 288	13. 490 - 139	14. 174 - 149	15. 196 - 100	16. 398 - 114
--------	--------	--------	--------	--------	--------	--------	--------

17. 955 - 115	18. 363 - 227	19. 175 - 115	20. 423 - 176	21. 518 - 278	22. 130 - 112	23. 102 - 102	24. 791 - 518
--------	--------	--------	--------	--------	--------	--------	--------

25. 224 - 121	26. 393 - 355	27. 604 - 496	28. 607 - 326	29. 250 - 181	30. 659 - 380	31. 609 - 601	32. 895 - 265
--------	--------	--------	--------	--------	--------	--------	--------

33. 248 - 215	34. 866 - 595	35. 787 - 655	36. 872 - 265	37. 758 - 524	38. 947 - 637	39. 586 - 165	40. 456 - 326
--------	--------	--------	--------	--------	--------	--------	--------

41. 397 - 241	42. 567 - 536	43. 710 - 380	44. 537 - 172	45. 508 - 445	46. 386 - 256	47. 617 - 544	48. 703 - 391
--------	--------	--------	--------	--------	--------	--------	--------

Name:.................................   Date:...............................

Find the difference

Complete all the activities (Subtraction)

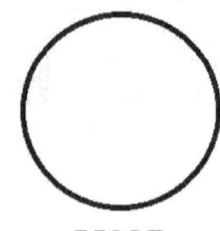

SCORE

1. 118 − 112 = 6	2. 942 − 147 = 795	3. 640 − 450 = 190	4. 676 − 670 = 6	5. 291 − 195 = 96	6. 843 − 697 = 146	7. 697 − 134 = 563	8. 362 − 347 = 15
9. 469 − 134 = 335	10. 893 − 423 = 470	11. 153 − 130 = 23	12. 572 − 288 = 284	13. 490 − 139 = 351	14. 174 − 149 = 25	15. 196 − 100 = 96	16. 398 − 114 = 284
17. 955 − 115 = 840	18. 363 − 227 = 136	19. 175 − 115 = 60	20. 423 − 176 = 247	21. 518 − 278 = 240	22. 130 − 112 = 18	23. 102 − 102 = 0	24. 791 − 518 = 273
25. 224 − 121 = 103	26. 393 − 355 = 38	27. 604 − 496 = 108	28. 607 − 326 = 281	29. 250 − 181 = 69	30. 659 − 380 = 279	31. 609 − 601 = 8	32. 895 − 265 = 630
33. 248 − 215 = 33	34. 866 − 595 = 271	35. 787 − 655 = 132	36. 872 − 265 = 607	37. 758 − 524 = 234	38. 947 − 637 = 310	39. 586 − 165 = 421	40. 456 − 326 = 130
41. 397 − 241 = 156	42. 567 − 536 = 31	43. 710 − 380 = 330	44. 537 − 172 = 365	45. 508 − 445 = 63	46. 386 − 256 = 130	47. 617 − 544 = 73	48. 703 − 391 = 312

Find the difference

Complete all the activities (Subtraction)

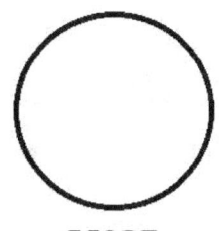

SCORE

1.	587 - 502 --------	2.	303 - 145 --------	3.	951 - 205 --------	4.	856 - 350 --------	5.	381 - 132 --------	6.	225 - 214 --------	7.	859 - 806 --------	8.	914 - 126 --------

| 9. | 229
- 227
-------- | 10. | 253
- 114
-------- | 11. | 517
- 496
-------- | 12. | 576
- 436
-------- | 13. | 363
- 281
-------- | 14. | 272
- 258
-------- | 15. | 540
- 239
-------- | 16. | 674
- 602
-------- |

| 17. | 976
- 503
-------- | 18. | 183
- 159
-------- | 19. | 917
- 242
-------- | 20. | 720
- 229
-------- | 21. | 671
- 208
-------- | 22. | 659
- 416
-------- | 23. | 388
- 265
-------- | 24. | 388
- 251
-------- |

| 25. | 946
- 342
-------- | 26. | 460
- 312
-------- | 27. | 718
- 459
-------- | 28. | 116
- 109
-------- | 29. | 366
- 239
-------- | 30. | 936
- 317
-------- | 31. | 937
- 872
-------- | 32. | 856
- 854
-------- |

| 33. | 527
- 213
-------- | 34. | 946
- 425
-------- | 35. | 853
- 243
-------- | 36. | 211
- 208
-------- | 37. | 366
- 356
-------- | 38. | 509
- 212
-------- | 39. | 395
- 385
-------- | 40. | 446
- 350
-------- |

| 41. | 429
- 320
-------- | 42. | 979
- 279
-------- | 43. | 922
- 140
-------- | 44. | 267
- 166
-------- | 45. | 497
- 298
-------- | 46. | 478
- 247
-------- | 47. | 319
- 194
-------- | 48. | 348
- 277
-------- |

Find the difference

Complete all the activities (Subtraction)

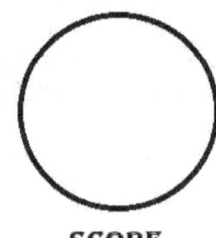

SCORE

1. 587 - 502 85	2. 303 - 145 158	3. 951 - 205 746	4. 856 - 350 506	5. 381 - 132 249	6. 225 - 214 11	7. 859 - 806 53	8. 914 - 126 788
9. 229 - 227 2	10. 253 - 114 139	11. 517 - 496 21	12. 576 - 436 140	13. 363 - 281 82	14. 272 - 258 14	15. 540 - 239 301	16. 674 - 602 72
17. 976 - 503 473	18. 183 - 159 24	19. 917 - 242 675	20. 720 - 229 491	21. 671 - 208 463	22. 659 - 416 243	23. 388 - 265 123	24. 388 - 251 137
25. 946 - 342 604	26. 460 - 312 148	27. 718 - 459 259	28. 116 - 109 7	29. 366 - 239 127	30. 936 - 317 619	31. 937 - 872 65	32. 856 - 854 2
33. 527 - 213 314	34. 946 - 425 521	35. 853 - 243 610	36. 211 - 208 3	37. 366 - 356 10	38. 509 - 212 297	39. 395 - 385 10	40. 446 - 350 96
41. 429 - 320 109	42. 979 - 279 700	43. 922 - 140 782	44. 267 - 166 101	45. 497 - 298 199	46. 478 - 247 231	47. 319 - 194 125	48. 348 - 277 71

Name:.................................   Date:................................

Find the difference

Complete all the activities (Subtraction)

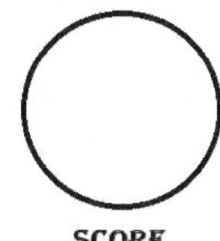

SCORE

1. 413	2. 181	3. 667	4. 717	5. 246	6. 364	7. 214	8. 338
- 295	- 104	- 631	- 563	- 163	- 291	- 185	- 179

9. 222	10. 902	11. 949	12. 175	13. 171	14. 947	15. 792	16. 753
- 125	- 312	- 641	- 132	- 139	- 387	- 609	- 665

17. 717	18. 932	19. 886	20. 182	21. 982	22. 137	23. 779	24. 340
- 558	- 686	- 360	- 135	- 301	- 118	- 413	- 268

25. 311	26. 327	27. 584	28. 791	29. 586	30. 976	31. 994	32. 982
- 119	- 108	- 106	- 592	- 423	- 173	- 411	- 476

33. 733	34. 135	35. 835	36. 511	37. 205	38. 355	39. 635	40. 702
- 371	- 113	- 512	- 431	- 149	- 283	- 420	- 472

41. 352	42. 193	43. 535	44. 394	45. 975	46. 247	47. 420	48. 492
- 162	- 163	- 400	- 298	- 288	- 131	- 103	- 348

Find the difference

Complete all the activities (Subtraction)

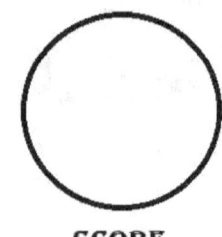

SCORE

| 1. | 413
- 295
118 | 2. | 181
- 104
77 | 3. | 667
- 631
36 | 4. | 717
- 563
154 | 5. | 246
- 163
83 | 6. | 364
- 291
73 | 7. | 214
- 185
29 | 8. | 338
- 179
159 |

| 9. | 222
- 125
97 | 10. | 902
- 312
590 | 11. | 949
- 641
308 | 12. | 175
- 132
43 | 13. | 171
- 139
32 | 14. | 947
- 387
560 | 15. | 792
- 609
183 | 16. | 753
- 665
88 |

| 17. | 717
- 558
159 | 18. | 932
- 686
246 | 19. | 886
- 360
526 | 20. | 182
- 135
47 | 21. | 982
- 301
681 | 22. | 137
- 118
19 | 23. | 779
- 413
366 | 24. | 340
- 268
72 |

| 25. | 311
- 119
192 | 26. | 327
- 108
219 | 27. | 584
- 106
478 | 28. | 791
- 592
199 | 29. | 586
- 423
163 | 30. | 976
- 173
803 | 31. | 994
- 411
583 | 32. | 982
- 476
506 |

| 33. | 733
- 371
362 | 34. | 135
- 113
22 | 35. | 835
- 512
323 | 36. | 511
- 431
80 | 37. | 205
- 149
56 | 38. | 355
- 283
72 | 39. | 635
- 420
215 | 40. | 702
- 472
230 |

| 41. | 352
- 162
190 | 42. | 193
- 163
30 | 43. | 535
- 400
135 | 44. | 394
- 298
96 | 45. | 975
- 288
687 | 46. | 247
- 131
116 | 47. | 420
- 103
317 | 48. | 492
- 348
144 |